Mechanics of Groundwater in Porous Media

Muhammad I. Haque

CRC Press is an imprint of the
Taylor & Francis Group, an **informa** business
A SPON PRESS BOOK

CRC Press
Taylor & Francis Group
6000 Broken Sound Parkway NW, Suite 300
Boca Raton, FL 33487-2742

First published in paperback 2017

Version Date: 20140428

ISBN 13: 978-1-138-07222-0 (pbk)
ISBN 13: 978-1-4665-8504-1 (hbk)

Library of Congress Cataloging-in-Publication Data

Haque, Muhammad I.
Mechanics of groundwater in porous media / Muhammad I Haque.
pages cm
Summary: "The present text evolved from the lectures delivered by the author to senior undergraduates and graduate students, primarily in Civil Engineering Curriculum, during the past twenty-five years or so, at The George Washington University. During this period, many students have contributed, directly or indirectly, by raising thoughtful questions. To all of them I owe my indebtedness"-- Provided by publisher.
Includes bibliographical references and index.
ISBN 978-1-4665-8504-1 (hardback)
1. Groundwater flow--Mathematical models. 2. Porosity--Mathematical models. 3. Aquifers. I. Title.

TC176.H38 2014
551.49--dc23 2014004187

Visit the Taylor & Francis Web site at
http://www.taylorandfrancis.com

and the CRC Press Web site at
http://www.crcpress.com

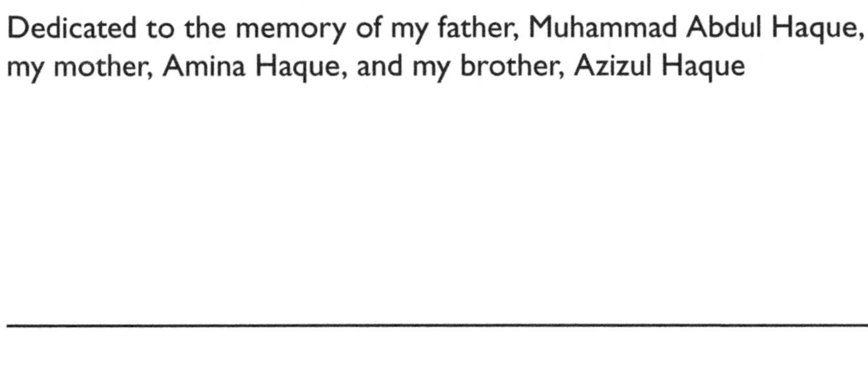

Dedicated to the memory of my father, Muhammad Abdul Haque, my mother, Amina Haque, and my brother, Azizul Haque

Contents

Preface

This book evolved from the lectures that I delivered to senior undergraduate and graduate students during the past 25 years or so, primarily in the civil engineering curriculum at The George Washington University. During this period, many students have contributed directly or indirectly by raising thoughtful questions. I am indebted to all of them.

There are a number of textbooks that treat the subject of groundwater either from a qualitative or from a rather abstract mathematical perspective. In this book, I have tried to maintain a balance between the two viewpoints. Nevertheless, a working knowledge of university level mathematics generally obtained during the first two years in an engineering curriculum, or a qualitative science field, with some exposure to geology will provide an adequate preparation on the part of the reader. Mathematical complexity, if any, is gradually introduced throughout the text. I have, however, made liberal use of graphical illustrations to aid in the comprehension of physics and mathematical concepts, as applied to the field of groundwater hydrology. This aspect, coupled with a number of completely worked out illustrative problems in the text, should appeal both to students and practicing engineers.

This book addresses the subject of steady-state motion of groundwater in porous media only. An exception is the application of Theis' transient flow equations to the problem of the determination of aquifer characteristics in Chapter 7. Although topics such as unsteady flows in unsaturated media with possible application to the transport of contaminants constitute an important aspect of groundwater mechanics, these are not treated here to limit the volume of the book.

I express my gratitude to my colleague and friend, Professor Khalid Mahmood, with whom I spent hours discussing various topics ranging from the arts to the sciences. Without his constant inspiration and encouragement, this task would not have been successfully completed. I have immensely benefitted from previous authors, in particular Professor M. E. Harr and Professor Otto D. L. Strack, during the preparation of this manuscript. A special note of thanks is due to Professor Erik Thompson of

Colorado State University for introducing the finite element method to me. To all of them, I express my gratitude.

I take this opportunity to thank the members of the Spon Press (imprint of Taylor & Francis Group), especially Tony Moore and Amber Donley, for their unwavering support during the entire process of publication.

Finally, I express my gratitude to my children, Sameera, Kashif, and Omar for their help in the preparation of the manuscript. As always, I thank my wife, Tahira, for understanding and support.

Author

Dr. Muhammad I. Haque currently serves as a professor of engineering and applied science in the Department of Civil and Environmental Engineering, School of Engineering and Applied Science, at The George Washington University, where he has offered a course on groundwater and seepage to senior undergraduate and upper level graduate students for almost a quarter of a century. A recipient of graduate degrees in hydraulics and mechanics, he has spent over four decades in research and practice involving water resources engineering. His area of expertise includes the finite element method and its theory and application to incompressible materials, both solids and fluids.

Besides publishing numerous articles in technical journals, Dr. Haque has contributed chapters to the *Encyclopedia of Fluid Mechanics* and to *Civil Engineering Practice 2/Hydraulics/Mechanics* and has coedited the book entitled *Mechanics of Alluvial Channels*. He is an excellent teacher and has received awards for excellence in teaching from the Engineering Council of the School of Engineering and Applied Science. He has also been recognized nationally as the Most Outstanding Civil Engineering Advisor by the American Society of Civil Engineers.

Chapter 1

Introduction

In this chapter, we introduce the overall picture of groundwater, its relationship with the hydrological cycle, and other terminology used in the mechanics of groundwater flow through porous media. The term *groundwater*, broadly speaking, refers to water that resides beneath the surface of the earth. However, to a groundwater hydrologist, or a geotechnical engineer, the term has a more restricted connotation: It implies a body of water that completely saturates the porous medium and the pressure at any point inside the body of water is equal to or above the atmospheric pressure. More on this aspect of water residing beneath the earth surface follows later in this chapter. We first look at some of the terminologies that are relevant for a further discussion. A water-bearing geological formation is called an *aquifer* if it stores enough water that can permeate through the geological formation under ordinary hydraulic gradients (or field conditions). Todd (1959) relates the term to its Latin roots—*aquifer*, combining form of *aqua* (water), and *ferre* (to bear). Thus, aquifer literally means water bearing. The important aspect of this definition is the fact that enough water can be economically mined from an aquifer. There are three more terms used in the definition of geological formation with regard to groundwater and its transmission. These are *aquiclude*, *aquitard*, and *aquifuge*. All these terms have their roots in Latin, according to Todd. An *aquiclude* is a geological formation that may contain a significant amount of water but is incapable of transmitting it under ordinary field conditions. A typical example is a clay layer. For all practical purposes, an *aquiclude* represents an impermeable formation. It generally forms the confining layer of a confined aquifer (to be discussed subsequently). An *aquitard* refers to a geological formation that transmits water at a very low rate compared with an aquifer. It often acts as a leaky formation between two aquifers. The term *aquifuge* on the other hand refers to a geological formation that neither contains nor transmits water. Typical examples include unfractured igneous rocks such as granite.

All rock masses contain *solid skeleton* and *void spaces*. These void spaces are referred to by different designations; for instance, interstices, pores, or pore spaces are synonymous with the word *void spaces*. The rock properties that affect the storage and movement of groundwater include *porosity* and *coefficient of permeability*. These terms are defined in quantitative terms in Chapters 2 and 3. However, a quasi-quantitative definition will suffice here for the time being. It is the porous space in an aquifer that stores water. Thus, greater the porosity of an aquifer, greater its capacity to store fluids. It is this property of the rock mass that is described quantitatively in the definition of the term *porosity*. Likewise, more permeable the aquifer, greater the ease with which the aquifer can transmit water. This quality of aquifer is described more precisely in a quantitative manner in the definition of the term *coefficient of permeability* in Chapter 3.

In a given specimen of aquifer, not all the pore space is necessarily filled with water. If the void space is completely filled with water, the *degree of saturation* (or *saturation*) of the aquifer is referred to as 100%. If only one-half of the void space is filled with water, the degree of saturation of aquifer is said to be 50%. Thus, the degree of saturation indicates the proportion of void space that is occupied by water. Its precise mathematical definition would be the ratio of volume of water, V_w, to the volume of void space, V_v, of the soil sample, that is, the degree of saturation, $S_w = V_w/V_v$. Sometimes, this ratio is expressed simply as decimal fraction and other times as a percentile ratio, as described earlier. Thus, greater the degree of saturation, wetter the soil is. There is another term that is similar (but not identical) to the degree of saturation, which also describes the wetness of soil sample. It is designated as the *moisture content* (or *water content*) of the soil. It is defined as the ratio of volume of water, V_w, contained in a soil sample to the total (nominal) volume, V, of the soil sample, that is, the moisture content, $\theta = V_w/V$. Sometimes, it is also defined as the gravimetric, instead of volumetric, ratio.

1.1 HYDROLOGICAL CYCLE

A schematic sketch of the hydrological cycle is shown in Figure 1.1. In this sketch, the scale toward the earth crust is highly exaggerated in comparison with the mean earth radius. This exaggeration is needed to highlight the essential features of the hydrological cycle, such as the migration of water vapor after evaporation. It should be stated at the very outset that almost all water vapors (including clouds), residing in the atmosphere, occupy only the first layer of the atmosphere, close to the earth surface, commonly called the *troposphere*. The thickness of the troposphere is considered to be about 12 km, which is much smaller than the average radius (6370 km)

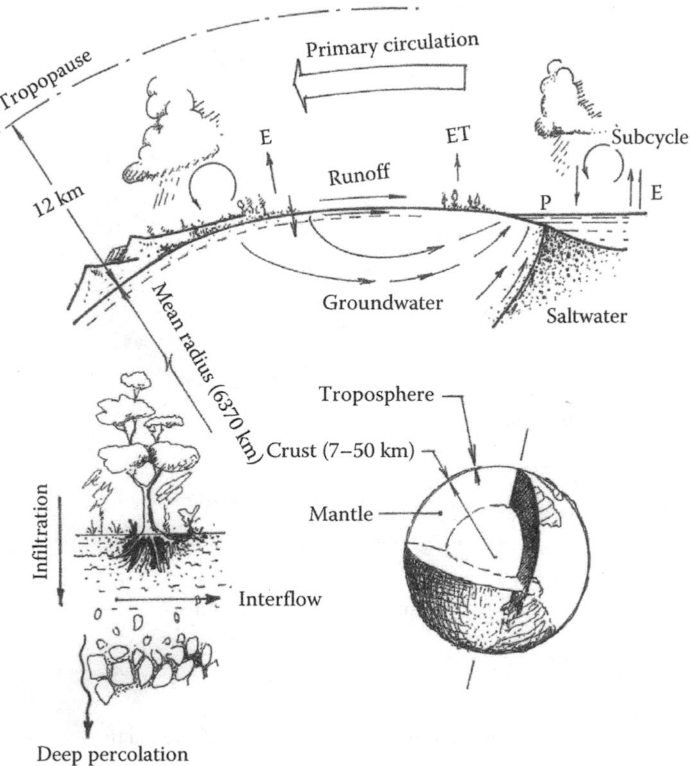

Figure 1.1 A schematic view of the hydrological cycle. *Note:* E, evaporation; ET, evapotranspiration; P, precipitation.

of the earth. It is therefore necessary to exaggerate the scale near the earth surface to cover the essential details without losing sight of the fact that the hydrological cycle covers a space comparable to a significant part of the earth surface. Looking from another perspective, the hydrological cycle is essentially a global phenomenon.

As the term *hydrological cycle* implies that it is a cycle in which any point (or event) can be viewed as the starting point; thereafter, conceptually speaking, the cycle perpetuates *ad infinitum*. Although it is arbitrary, precipitation is generally considered as the starting point (or event). During the hydrological cycle, the water from land, seas, and vegetation is evaporated; the vapor is transported through the atmosphere and eventually, under the right physical conditions, the vapor condenses into precipitation. Thus completing the cycle! Within this cycle, there are a number of subcycles, as indicated in Figure 1.1. The groundwater constitutes a part of hydrological cycle.

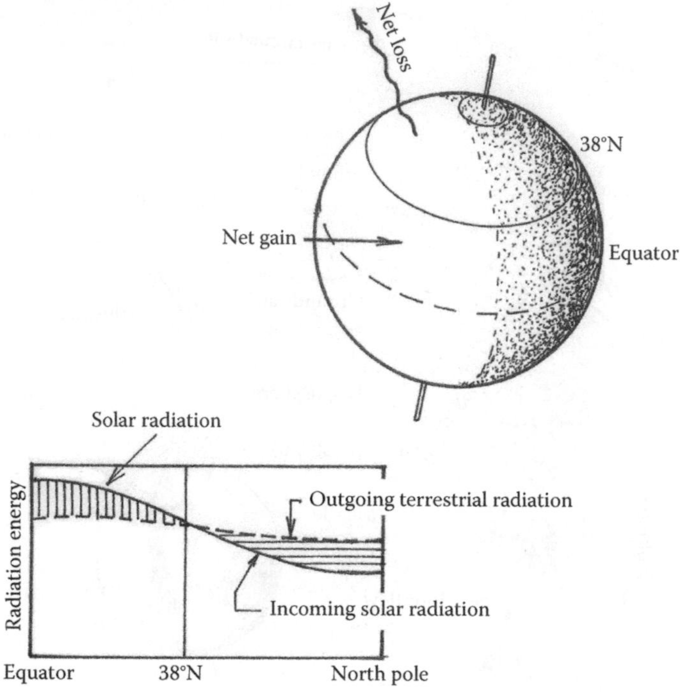

Figure 1.2 Net annual gain and loss of energies by parts of the earth.

In the hydrological cycle, there are three fundamental processes involved. These are *precipitation*, *evaporation of water* from the seas and land masses, *migration of vapors through the atmosphere*, and the eventual condensation of water vapors. What does keep this cycle going? The short and quick answer is that the cycle is powered by the solar energy. To understand how the solar radiation helps maintain the cycle, Figure 1.2 has been prepared. It shows that in the northern hemisphere north of latitude 38° N, there is a *net loss* of radiant energy annually—this part receives less solar energy than it radiates out to the space, while below 38° N toward the equator there is a *net gain* of the radiant energy annually by the earth. This average annual energy imbalance, between parts of the earth surface, cannot be maintained year after year; the system must eventually come to a stable equilibrium state. As an effort to achieve stability on such a large scale, the system sets up convection currents both in the atmosphere and the oceans of the earth. Although energy transfer by thermal conduction through the earth crust plays some role in achieving equilibrium, it is insignificant compared with the convective transfer of energy.

There is another facet of the cycle worth looking at. This has to do with the average annual imbalance between evaporation from and precipitation to the oceanic surfaces of the world. The ocean surfaces evaporate more water annually than they receive through direct precipitation. This imbalance of water mass is, however, mitigated by the surface and subsurface migration of water from the continents to the oceans.

Finally, a few words about the distribution of the world's free water! Not all the water of earth is free. An undetermined amount is held by chemical bonds in the minerals of the earth. Of the free water, 97% is contained by oceans. The next large storage of water is provided by glaciers and polar ice caps. After glaciers and polar ice caps, the groundwater stores about 0.61% of the earth's free water. The remaining amount is stored collectively in lakes, ponds, rivers, and soil moisture. On the average, only 0.001% of free water resides in the atmosphere at any time. Despite this low amount of vapor present in the atmosphere, it is this that is responsible for a large amount of annual precipitation, because of the rapidity with which water vapor enters and leaves the atmosphere. To account for the global-average precipitation, the entire atmospheric vapor must be replaced 40 times a year. Or from another point of view, the average *residence time* of atmospheric moisture is slightly over 9 days (Eagleson, 1970).

1.1.1 Noncyclic water

Some of the water remains dormant for a long time and does not participate on a regular basis in the hydrological cycle. These waters are collectively classified as noncyclic waters. Among these, the *juvenile water* participates for the first time in the hydrological cycle. The juvenile water could be from three different sources: (1) magmatic, (2) volcanic, or (3) cosmic sources. Of the noncyclic water, the term *fossil water* refers to the water that is buried at the time of deposition of sediments during rock formation and remains in isolation for a long geological period. The term *connate water* refers to the water that remains in isolation for a long duration.

1.2 VERTICAL MOISTURE PROFILE

The part of precipitation that enters the earth surface is called the *infiltration*. Of this, a part called *interflow* moves parallel to the ground surface through relatively more pervious soil structure (inset Figure 1.1). The remaining part redistributes and eventually percolates downward due to the action of gravity until it merges into a saturated reservoir of water resting on the impervious stratum. This redistribution and downward migration of infiltrated water causes vertical moisture profile (Figure 1.3b). Broadly speaking, the vertical moisture profile can be classified into two zones: one called the zone of aeration, close to the earth surface, and the other called the zone of saturation, away from the earth surface. The zone of aeration

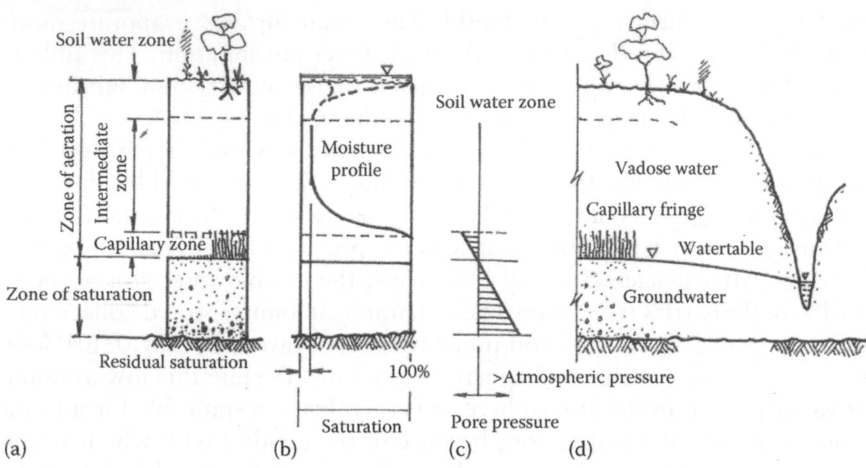

Figure 1.3 Definition sketch for the vertical moisture profile: (a) classification of moisture zones, (b) moisture profile, (c) vertical variation of pore pressure, and (d) a schematic cross section of aquifer.

is further classified into three categories: the *soil water zone*, *intermediate zone*, and the *capillary zone*. Collectively, the water contained in the zone of aeration is also referred to as the *vadose* water (Figure 1.3d). The main characteristic of vadose water is the fact that it only represents a part of the three-phase system: water, gas (generally air), and solid. Furthermore, the pore pressure (the pressure inside the water occupying the pore space) of vadose water is less than, or equal to, the atmospheric pressure. *The surface where the pore pressure is atmospheric is called the watertable or the phreatic surface.* In Figure 1.3, the lower edge of the *capillary zone* represents the *watertable*, where the pore pressure is atmospheric. The subclassification of the vadose zone is further described in the following text.

1.2.1 Soil water zone

The soil water zone starts from the ground surface and extends downward through the major root systems of the trees. The moisture content in this zone depends largely on the availability of water at the ground surface. Temporarily, the soil in this zone can become completely saturated after a significant rain storm, or surface irrigation.

1.2.2 Intermediate zone

It starts from the lower edge of the soil water zone and ends at the upper limit of the capillary zone. Its thickness (or depth) depends largely on the location of the watertable (or phreatic surface). For deep watertable conditions, the thickness of intermediate zone can approach several hundred feet. Its primary role is to provide vertical seepage of excess (gravity) water, under the influence of gravitational pull. The water that is held up by the soil particles against the gravitational pull is called the *pellicular* water. The hygroscopic and capillary forces are instrumental in attaching this water to the surface of soil particles.

1.2.3 Capillary zone

The region between the intermediate zone and groundwater is called the capillary zone. The water in the capillary zone is held in interstices by the same forces that cause capillarity in narrow tubes in physical laboratories. An important feature of capillary zone is the fact that despite saturation, the pressure in pore water is subatmospheric. In other words, if the atmospheric pressure is taken as the zero of pressure scale, the pore pressure in the capillary zone becomes negative.

The moisture profile in the zone of aeration, in general, and that in the capillary zone, in particular, is dependent on the grading of aquifer material. Figure 1.4 highlights the dependence of moisture profile on the gradation of the aquifer material. An ensemble of aquifer material is said to be *well graded* (or *poorly sorted*) if it represents a variety of grain sizes; it is called poorly graded if the ensemble represents the predominance of a single grain size. It can be seen from Figure 1.4 that the moisture decreases rapidly toward residual saturation with height, z, in the case of poorly graded soil. On the other hand, in the case of well-graded soils, the moisture profile gradually decreases with height. This behavior is attributed to

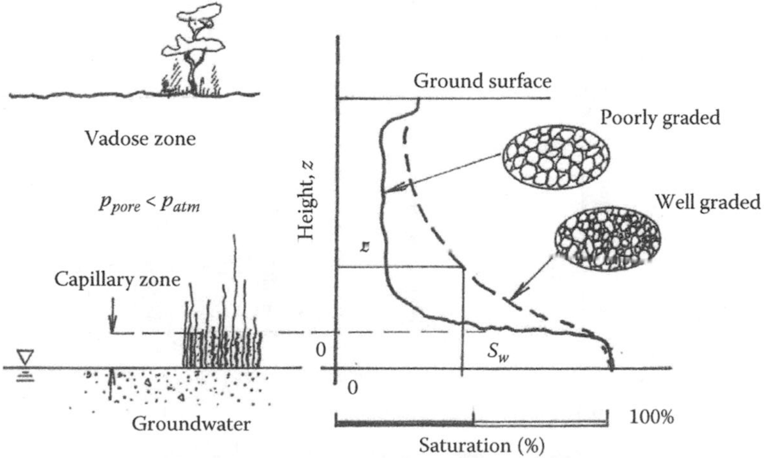

Figure 1.4 The influence of gradation on moisture profile.

the geometric differences in interstitial spaces between solid grains of the well-graded and poorly graded aquifers.

In order to understand the physical mechanism operative in the development of moisture profile in the capillary zone, it is instructive to replace conceptually the interstitial space of the aquifer by an idealized model of the void space. For this purpose, we replace the tortuous interstitial space between the solid grains of the aquifer by the cylindrical tubular void spaces of an ensemble of capillary tubes of different inner radii. Each capillary tube is oriented vertically in space, and its radius is supposed to represent, in some sense, the mean radius of a twisted interstitial path. The rise of water due to capillary action in such an idealized ensemble of tubes can be easily determined from the basic laws of fluid mechanics. This capillary action is exemplified in Figure 1.5. The abscissa in this figure represents the normalized radius, r/r_0, of the capillary tube, varying continuously from zero to a maximum value of $r/r_0 = 1.0$. The corresponding ordinate represents the normalized capillary rise h_c/h_{c0} in a tube of inner radius r. The curve in this figure is based on the fact that the capillary rise, h_c, in a tube is inversely proportional to the inner radius, r, of the tube.

At this stage, we introduce another simplification in order to make the problem even more tractable. We assume the following simplifying assumption:

$$N_r \propto \frac{1}{r^2} \tag{1.1}$$

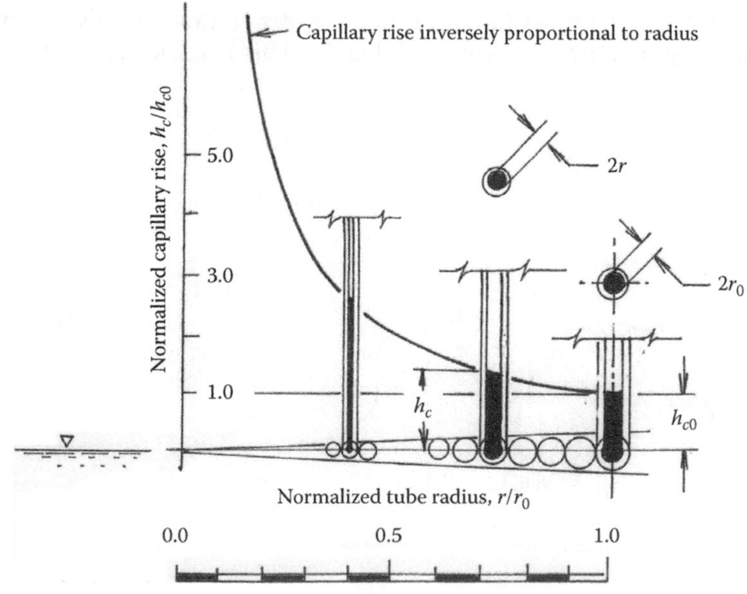

Figure 1.5 Capillary rise in an ensemble of tubes.

where N_r represents the total number of capillary tubes in the ensemble with the inner radii equal to r. In other words, the number of tubes increases four times as the radius reduces to one-half. This assumption is more in keeping with the well-graded aquifer than with the poorly graded ones. More importantly, this simplification ensures that *each group of capillary tubes with the same radius* covers the same proportion of the total cross-sectional area represented by the ensemble of capillary tubes. Thus, it can be shown (Appendix A) that the graph h_c/h_{c0} versus r/r_0 (Figure 1.5) is identical with the graph between z/z_f and the degree of saturation $S_w = V_w/V_v$, as shown in Figure 1.6.

Finally, as the size of the capillary tubes approaches zero, the forces that cause capillarity are dominated by those forces that cause the affinity of water molecules to adhere to the surface of mineral particles. Water

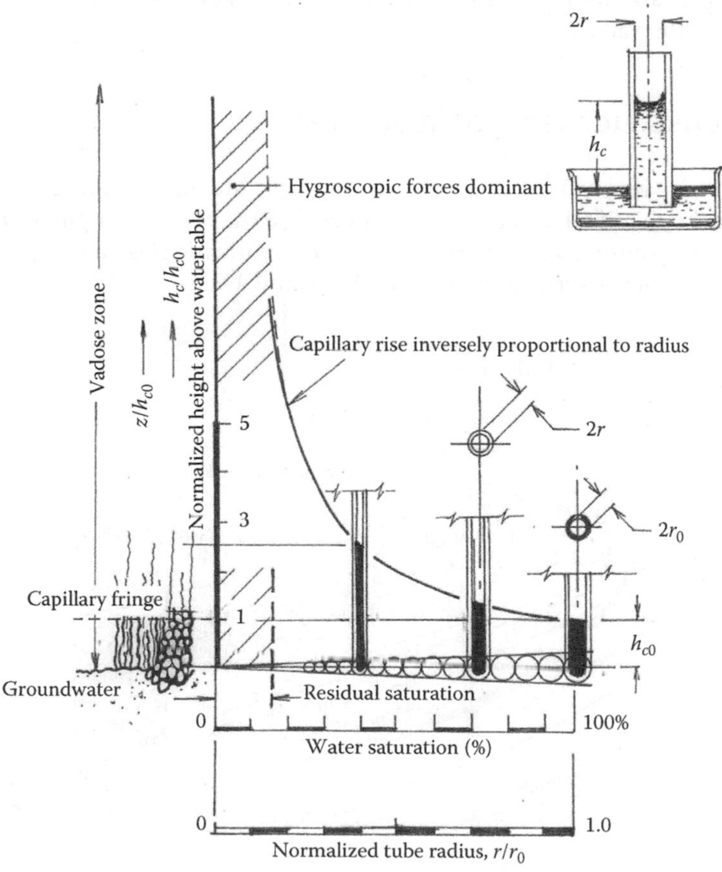

Figure 1.6 Moisture profile based on a simplified ensemble of capillary tubes.

held at or near the surface of mineral particles is called hygroscopic water (Polubarinova-Kochina, 1962). In the context of groundwater hydrology, the term *pellicular water* is used to embrace all kinds of waters that do not readily move under the influence of gravitational pull. The cross-hatching in Figure 1.6 implies the pore spaces where the hygroscopic and pellicular waters dominate the capillary action. Thus, for narrower interstitial space, this leaves the aquifer with the residual saturation as shown in the figure.

Finally, as closing remarks, it should be noted that the physics of the movement of water through twisted, uneven interstitial spaces is quite different from the physics that determines the capillary rise. For instance, the physics of capillary rise through a constant tubular passage can never replicate the *hysteresis effects* normally associated with moisture curve. The analysis in the forgoing is intended to demonstrate that simple physical principles can contribute to the fundamental understanding of moisture profile in the vadose zone.

1.3 CLASSIFICATION OF AQUIFERS

Aquifers are generally of extensive areal extent. Broadly speaking, the aquifers can be classified into two main categories: the confined aquifer and the unconfined aquifer. These two types are schematically shown in Figure 1.7, and are further described in Sections 1.3.1 and 1.3.2.

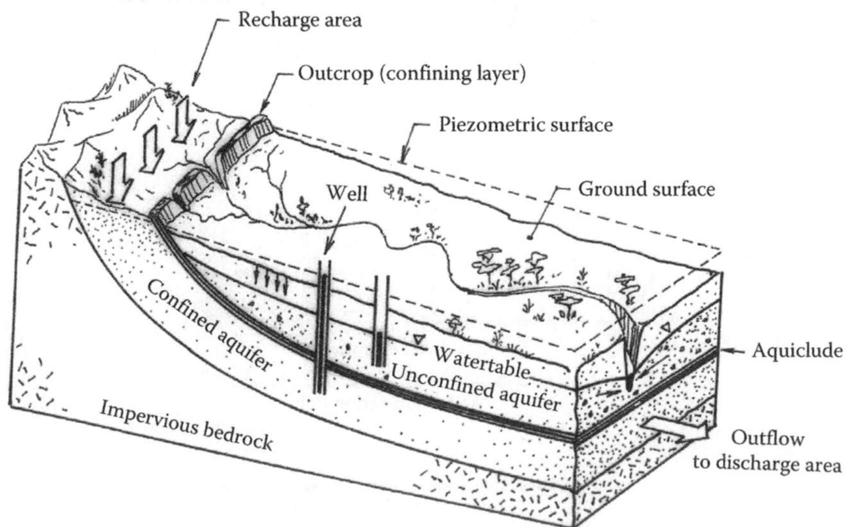

Figure 1.7 A schematic sketch for the classification of aquifers.

1.3.1 Confined aquifer

An aquifer that is confined from above by a layer of aquiclude, and also confined from below by either an aquiclude or impervious bedrock, is called a confined aquifer. The pore pressure (the pressure inside the groundwater) in such an aquifer is quite above the atmospheric pressure. The confined aquifer mostly serves as a conduit for the transportation of groundwater from the recharge area to the natural, or man-made, discharge site (Figure 1.7). Since water is considered highly incompressible (compressibility, $\beta = 4.5 \times 10^{-10}$ m^2/N), even a small amount of addition (or subtraction) of water volume to the confined aquifer significantly raises the pore pressure throughout the aquifer.

1.3.2 Unconfined aquifer

An aquifer whose upper boundary is defined by watertable (or phreatic surface) is called an unconfined aquifer. The lower boundary of such an aquifer may be a confining layer, or impervious bedrock. In the case of an unconfined aquifer, the undulations in the phreatic surface (watertable) are largely due to addition or subtraction of water volume. In comparison with a confined aquifer, in the case of an unconfined aquifer, addition or subtraction of a small amount of water does not significantly affect the pore pressure. The unconfined aquifers receive recharge from deep percolation of infiltrated surface water. They discharge either naturally into ground surface, rivers, lakes, oceans or into man-made discharge locations such as wells.

1.3.3 Perched aquifer

The perched aquifers are schematically illustrated in Figure 1.8. These aquifers are essentially unconfined aquifers that rest on clay lenses of limited areal extent. The perched aquifers are dispersed in the unsaturated zone between the main watertable (of an extensive unconfined aquifer) and the ground surface, as shown in the figure.

1.4 RIVER–AQUIFER INTERACTION

The interactions between the rivers and the aquifers are illustrated in Figure 1.9. There are essentially two types of interactions: (1) either the stream is gaining (effluent stream) water from the aquifer (Figure 1.9a), or (2) it is losing (influent stream) water to the aquifer (Figure 1.9b and c). When the groundwater slopes toward the river bank, the river is gaining water; otherwise, it is losing water to the aquifer.

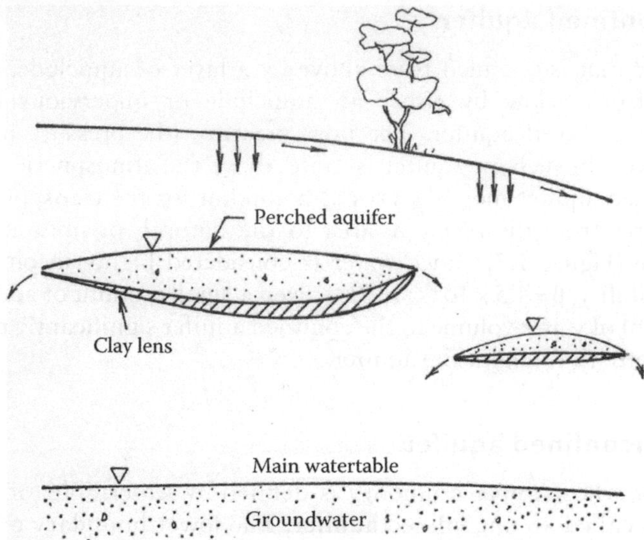

Figure 1.8 Perched aquifer.

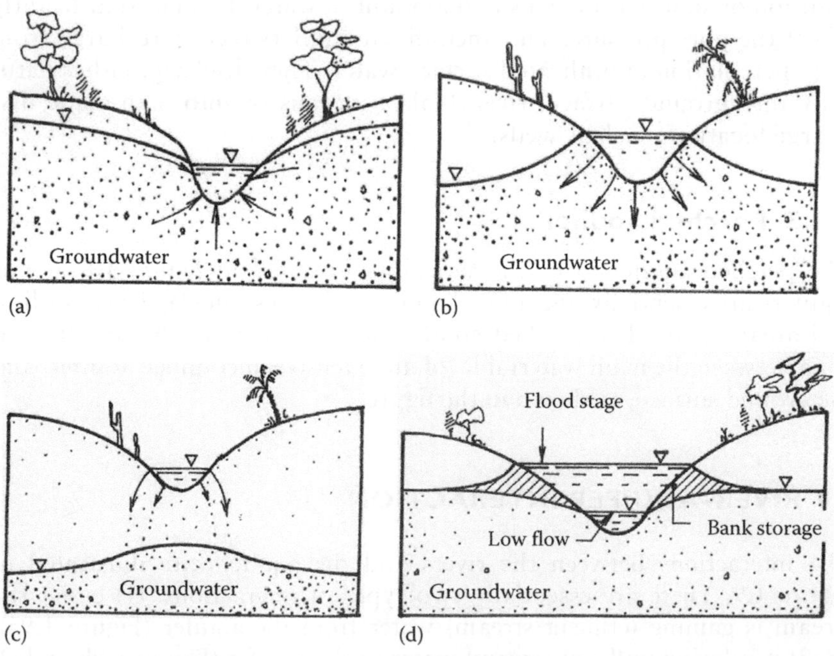

Figure 1.9 River–aquifer interaction: (a) effluent stream, (b) influent stream, (c) influent stream (deep watertable), and (d) bank storage.

The same river could be effluent along a reach of the river, while on the other reaches it could be influent. Even at the same reach, a river can act as the influent stream during high flood levels, and it may act as an effluent stream during low flood levels. The water stored in the aquifer during high and low river stages is termed as *bank storage* (Figure 1.9d). The water provided by the aquifer to the river is said to constitute the *base flow* of the river. It is the base flow that maintains discharge in the river during prolonged absence of precipitation in the catchment area.

1.5 HOMOGENEITY AND ISOTROPY OF AQUIFERS

There are two properties of aquifers that play a paramount role in the mechanics of groundwater through porous media. These are *homogeneity* and *isotropy* of aquifers with respect to some *physical quantity of interest*. In our case, the quantity of interest includes the coefficient of permeability and porosity, to mention a few. A medium of flow, or aquifer, is said to be homogeneous if the quantity of interest does not depend on the location of the point. In other words, the quantity of interest remains uniform throughout the medium. If a medium is not homogeneous, it is called heterogeneous. The medium of flow is said to be isotropic if, at a given point, the quantity of interest does not depend on the direction; otherwise, it is called anisotropic. To be concrete, we shall take the *coefficient of permeability* as the quantity of interest for further discussion. There are four possibilities with regard to this physical quantity and the aquifer: (1) the aquifer is homogeneously isotropic; (2) the aquifer is heterogeneously isotropic; (3) the aquifer is homogeneously anisotropic; and (4) the aquifer is heterogeneously anisotropic. These possibilities, or conditions, are graphically shown in Figure 1.10a through d.

For isotropic aquifers, the coefficient of permeability, K, at a given point of flow medium is represented graphically by the length of an arrow. The direction of the arrow indicates the direction along which the coefficient of permeability has been implied (or measured). For an isotropic aquifer, the length of the arrow does not change at a given point. Thus, the tip of the arrow traces a circular trajectory as its direction is varied. In the case of an anisotropic aquifer, however, the graphical illustration becomes more complex, because of the fact that the coefficient of permeability becomes a second-order tensor, as described in detail in Chapter 3. Nevertheless, the directional dependence of coefficient of permeability in the case of an anisotropic aquifer can be obtained from the so-called ellipse of direction. These ellipses are schematically shown for anisotropic media in Figure 1.10c and d. The semidiameters of these ellipses are proportional to the *square root of the directional coefficient*

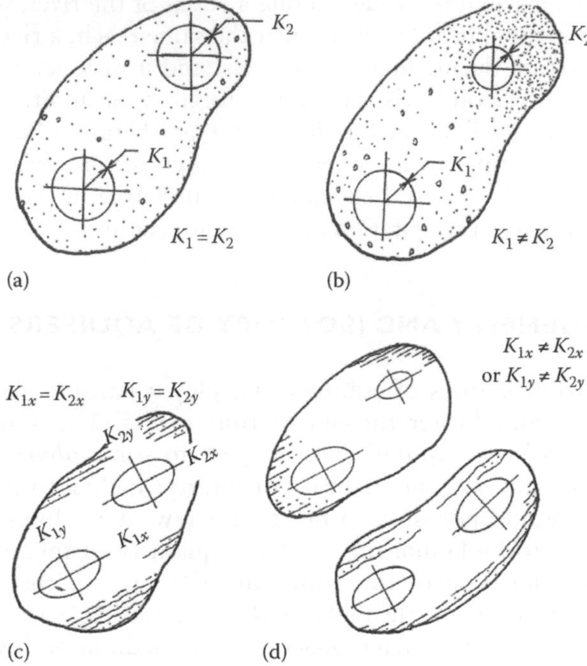

Figure 1.10 Types of homogeneity and isotropy of aquifers: (a) homogeneously isotropic, (b) heterogeneously isotropic, (c) homogeneously anisotropic, and (d) heterogeneously anisotropic.

of permeability. Furthermore, the directions of maximum and minimum diameters of the ellipse do represent the *principal directions* of anisotropy of the aquifer. Incidentally, these directions do correspond with the physical directions along and across the bedding planes of the aquifer.

Finally, in these sketches, if the size of the circle changes from point to point, the isotropic aquifer is considered heterogeneous. Likewise, if the shape or the orientation of the ellipse changes from point to point, the anisotropic aquifer is also considered heterogeneous.

1.6 ILLUSTRATIVE PROBLEMS

1.1 The phreatic surface (or watertable) levels with respect to an arbitrary datum are recorded on the map for three observation wells, A, B, and C, as shown in the following figure. Construct the contours of the watertable and find the direction of the steepest descend, using the data that have been provided on the map.

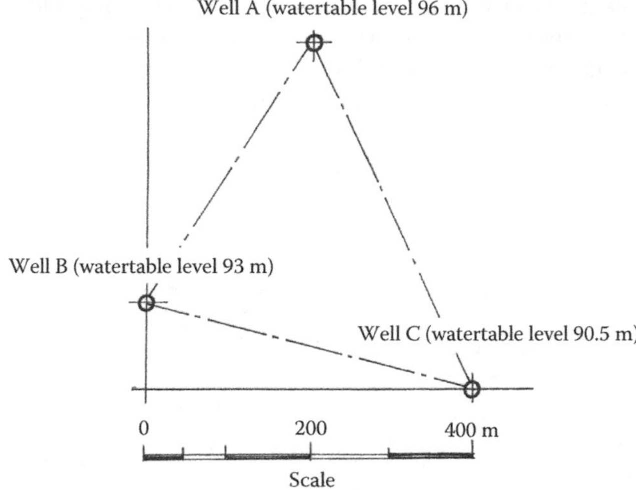

A contour map for the groundwater surface within the triangle ABC is constructed by linear interpolation. The groundwater contours and the line of steepest descend, which is at right-angle to the contour lines, are shown in the following figure. It may be emphasized here that the linear interpolation of the watertable, between three points ABC, represent a plane surface. The intersection of this plane surface with a horizontal surface yields a line that may be called *strike*, in keeping with the terminology used in structural geology. The strike lines are also shown in the figure.

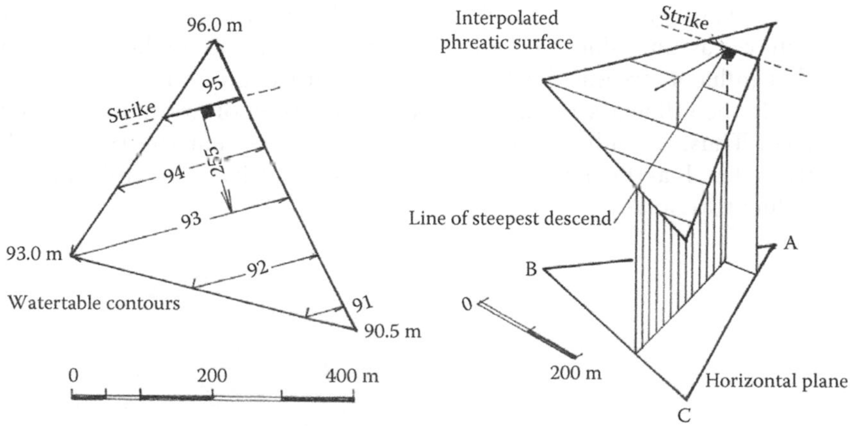

1.2 Groundwater contours are shown on the following map along with the course of a river. Determine the river reaches along which the stream behaves like an *effluent* and *influent* stream. Use the data that have been provided on the map for your analysis.

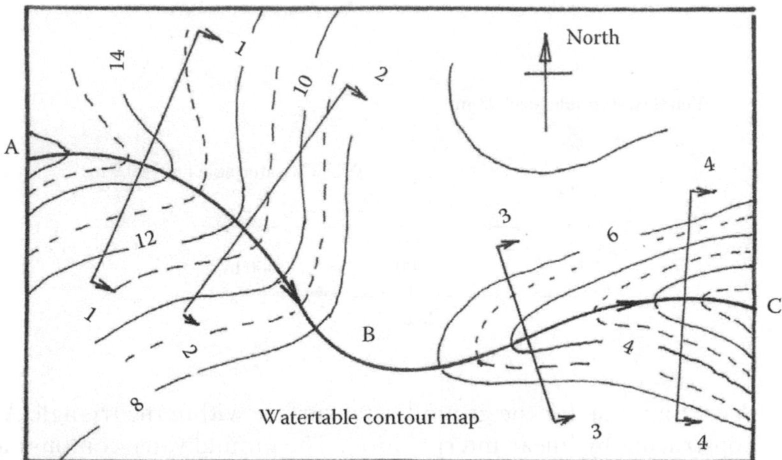

Watertable contour map

We construct typical cross sections of the phreatic surface across the river. Locations of these sections are identified on the map, and the cross sections themselves are shown in the following drawings. Cross sections, section 1-1 and section 2-2, show that the phreatic surface dips away from the river to the aquifer. Thus, along the reach from section 1-1 to section 2-2, the river behaves like an influent (loosing) stream. Likewise, sections 3-3 and 4-4 show that the phreatic surface slopes toward the river, indicating seepage from the aquifer into the river. Thus, along the reach between these two sections, the river behaves as an effluent (gaining) stream. It may be further observed that influent streams flow over the ridge of the groundwater surface, while the effluent streams flow along the valley of groundwater surface. Thus, a study of the contour map shows that the river reach from A to B and from B to C behaves, respectively, as the influent and effluent stream.

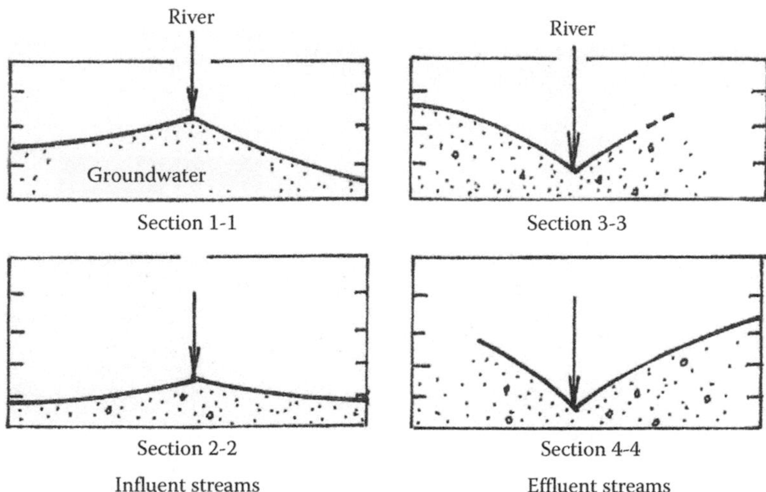

SUGGESTED READINGS

Related to hydrology and the mechanics of groundwater flow in general:

Bear, J. 1979. *Hydraulics of Groundwater*, McGraw-Hill, Inc., New York.

Bear, J. 1988. *Dynamics of Fluids in Porous Media*, Dover Publications, Inc., New York.

Bras, R. L. 1990. *Hydrology: An Introduction to Hydrologic Science*, Addison-Wesley, Reading, MA.

Eagleson, P. S. 1970. *Dynamic Hydrology*, McGraw-Hill, Inc., New York.

Fetter, C. W. 2001. *Applied Hydrogeology*, 4th edn., Prentice Hall, Inc., Upper Saddle River, NJ.

Freeze, R. A. and J. A. Cherry. 1979. *Groundwater*, Prentice Hall, Englewood Cliffs, NJ.

Polubarinova-Kochina, P. Ya. 1962. *Theory of Ground Water Movement*, translated from the Russian by J. M. Roger De Wiest, Princeton University Press, Princeton, NJ.

Todd, D. K. 1959. *Ground Water Hydrology*, John Wiley & Sons, Inc., New York.

SUGGESTED READINGS

Chapter 2

Preliminaries

In this chapter, we present a synopsis of basic definitions, concepts, and the fundamental principles of fluid mechanics and soil mechanics, which are necessary prerequisites for an adequate understanding of the subject matter treated in this book. To help facilitate comprehension on an elementary but firm level, a deliberate attempt is made to develop these ideas from the rudiments of physics; the principles of general applicability are deduced from geometric and physical reasoning, with a minimum use of the mathematical abstraction. For the understanding of the subsequent sections, however, a minimum proficiency in mathematical skills generally acquired during the first 2 years of differential and integral calculus at a university level is assumed. Likewise, some elementary exposure to vector analysis and matrix theory is highly desirable.

2.1 PRELIMINARIES FROM FLUID MECHANICS

2.1.1 Stress vector and its vector resolution into normal and shearing components

Let us consider a material body in a state of static equilibrium under the action of external forces, as shown in Figure 2.1a. These external forces can be classified into two categories: the first includes those forces that act directly on the surface of the body by physical contact, and the second includes those forces that act indirectly on the mass of the body without any material contact. The first category is designated as the *surface traction*, and the second as the *body force*. While surface traction includes, for instance, the reactive forces developed at the points of contact of the body with other material bodies, body force includes, for instance, the force exerted by the earth's gravitational field on the entire mass of the material body.

Let us perceive the body to be dissected into two parts A and B, by an imaginary plane *abcd*, as shown in Figure 2.1a. We assume that the body

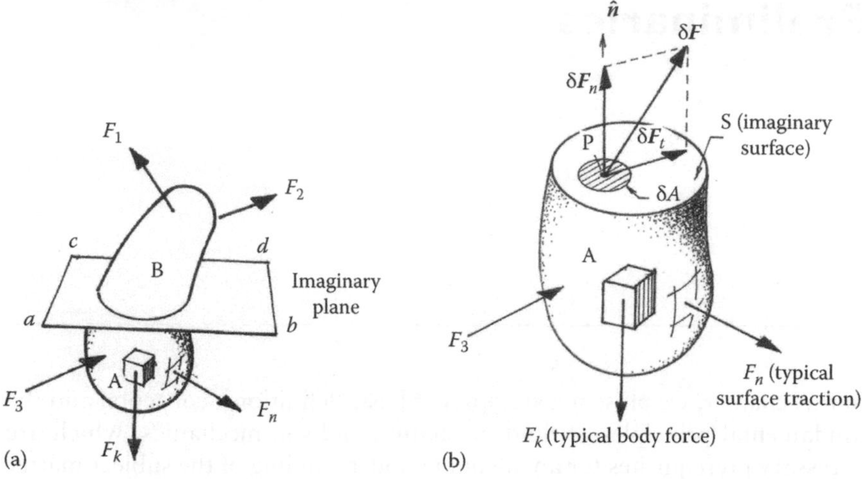

Figure 2.1 Stress vector and its vector resolution into normal and shearing vectors: (a) a material body dissected into parts A and B by an imaginary plane and (b) equilibrating forces on the imaginary surface S of part A.

as a whole, as well as any part thereof, is in a state of static equilibrium. Let us isolate the two parts from each other and focus our attention on the equilibrium of part A. Since, after dissection, the remaining external forces acting on part A are not necessarily in equilibrium, it becomes a logical necessity for us to assume the existence of some additional equilibrating forces acting on the imaginary surface S of part A; for, otherwise, part A cannot remain in equilibrium.

With regard to these equilibrating forces, we have no a priori knowledge of whether these forces act uniformly or act in a preferred direction over the imaginary surface S. Let us, therefore, consider, without loss of generality, a resultant elemental force δF acting at an oblique angle to the elemental area δA, as shown in Figure 2.1b. We now introduce the fundamental postulate of the mechanics of continuous media: *The ratio $\delta F/\delta A$ converges to a well-defined limit dF/dA at point P, and the moment due to δF about any point of δA vanishes, as δA tends to zero without leaving the point P.* In other words,

$$T^n \equiv \frac{dF}{dA} = \lim_{\delta A \to 0} \frac{\delta F}{\delta A} \tag{2.1}$$

where T^n is called the *stress vector (or traction)* and its direction is the limiting direction of δF. The magnitude and direction of the stress vector depend, in general, on the location of the point P and the orientation of the elemental area δA. The superscript n in the notation T^n emphasizes the dependence of the stress vector on the direction of the unit normal vector

$\hat{n}$ to the elemental area δA. The elemental force δF can be resolved into two orthogonal vector components δF_n and δF_t: the former acting in the direction of the unit outward normal vector to the elemental area δA, and the latter acting parallel to the plane of area δA, as shown in Figure 2.1b. It is convenient for future reference to introduce two additional limiting ratios:

$$\sigma \equiv \frac{dF_n}{dA} = \lim_{\delta A \to 0} \frac{\delta F_n}{\delta A} \tag{2.2a}$$

$$\tau \equiv \frac{dF_t}{dA} = \lim_{\delta A \to 0} \frac{\delta F_t}{\delta A} \tag{2.2b}$$

where

$\boldsymbol{\sigma}$ is the normal stress vector
$\boldsymbol{\tau}$ is the shearing stress vector

The normal stress vector is further qualified as the tensile stress vector, if δF_n acts in the direction of the outward unit normal vector, $\hat{\boldsymbol{n}}$, to the elemental area δA; otherwise, it is designated as the compressive stress vector. It may be noted that $\boldsymbol{\sigma}$ and $\boldsymbol{\tau}$ have the dimensions of force per area and thus represent the intensity—or, preferably, the density—of force distribution at a point. In the subsequent sections, we shall further use the concept of the normal and shearing stress vector in investigating the behavior of pressure in a fluid in a state of static equilibrium.

2.1.2 Static pressure at a point

By hypothesis, a liquid is a substance that yields under the action of a shear stress vector, however small the shear stress vector may be! As a corollary, the forces acting on the real or an imaginary surface of a liquid body at rest must act at right angle to the surface. Furthermore, since liquid cannot resist (significant) tensile stresses, these external surface forces must point toward the interior of the liquid body.

Let us now consider a body of liquid contained in a vessel to be at rest, as shown in Figure 2.2a. As before, we shall assume that the body as a whole, as well as any part thereof, is in static equilibrium. Let us further consider a cylindrical portion of the liquid (shaded cylinder in Figure 2.2a) removed from its surroundings as shown in Figure 2.2b. Such a body is called a free-body in mechanics. On this free-body, there again act two kinds of forces: one acts directly on the surface due to the action of the surrounding material, and the other indirectly on the mass due to the gravitational attraction of earth. In liquids that are in static equilibrium, the surface forces vary in general with area, but they always remain at a right angle to the surface.

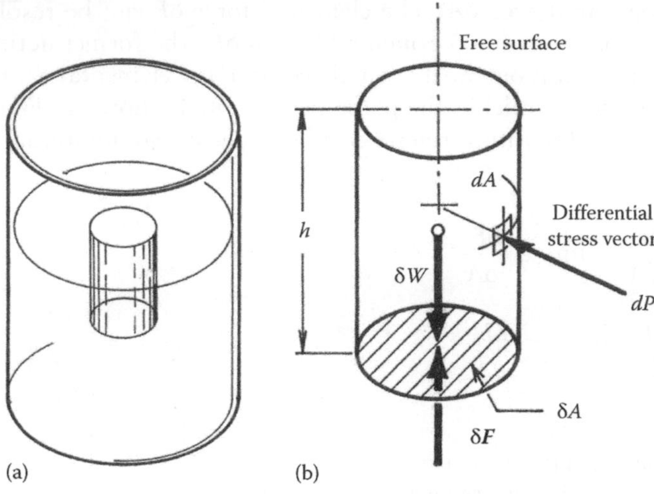

Figure 2.2 Definition sketch: (a) a liquid body at rest and (b) cylindrical free-body.

The body force, however, is proportional to the mass of the body, and it acts along the direction of acceleration due to gravity. For simplicity, we assume that at the top surface (free surface) there acts no force. At the bottom surface, however, there must act a resultant upward force δF, as shown in Figure 2.2b. On every differential area dA of the curved cylindrical surface acts some resultant differential force dP, which is oriented at a right angle to the surface. Since the free-body is in a state of static equilibrium, the algebraic sum of the components of forces along the vertical axis must equal zero.

Resolving forces along the vertical axis, and remembering that the differential forces on the cylindrical surface have no components along this axis, we conclude

$$\delta F = \delta W = \gamma \delta A h \tag{2.3}$$

where

- δW is the weight of the cylindrical free-body [MLT^{-2}]
- γ is the specific weight (weight per volume) of liquid [$ML^{-2}T^{-2}$]
- δA is the cross-sectional area of the cylinder [L^2]
- h is the height of cylinder [L]

The static pressure at any point P in a liquid, at a depth h from the free surface, is defined by the following limiting ratio:

$$p = \lim_{\delta A \to 0} \frac{\delta F}{\delta A} = \lim_{\delta A \to 0} \frac{\gamma \delta A h}{\delta A} = \gamma h \tag{2.4}$$

as δA contracts to zero without leaving the point P. *Thus, in a liquid with constant specific weight, the static pressure at any point depends only on the depth of the point from the free surface.*

2.1.2.1 The invariance of static pressure with direction

It is evident that the pressure, p, and the normal stress vector, σ, are similar concepts. Since normal stress vector at a given point depends in general on the orientation of the elemental area, it is natural to expect a similar behavior in the case of pressure. We are, therefore, led to the following question: How does the pressure change with the orientation of the elemental area δA, at a fixed point P? To answer this question, the illustration in Figure 2.3 has been prepared. This figure shows a cylindrical free-body isolated from the surrounding liquid in static equilibrium. Unlike the previous case, however, the bottom surface of this cylindrical free-body is inclined at an angle θ with respect to the horizontal plane. The centroid of the bottom (point P in Figure 2.3) is located at a depth h, from the free surface. Thus, the weight of the free-body can be obtained from the following equation:

$$\delta W = \gamma \delta A' h \tag{2.5}$$

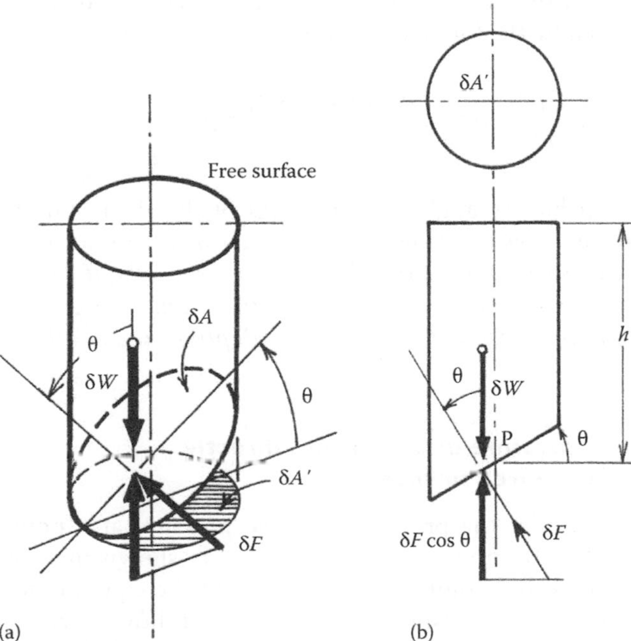

Figure 2.3 Invariance of static pressure with direction: (a) perspective view of the free body and (b) side view of the free body.

where $\delta A'$ represents the normal cross-sectional area of the cylindrical free-body. The normal cross-sectional area $\delta A'$ is, in fact, the orthogonal projection of the elliptical bottom area δA on the horizontal plane (Figure 2.3a). Thus, these elemental areas are related by the following equation:

$$\delta A = \frac{\delta A'}{\cos\theta} \tag{2.6}$$

On the elliptical bottom area δA, there must act a resultant force δF due to the surrounding liquid. This resultant force must act at right angle to the surface, and it should point toward the interior of the body, because of the hypothesis of liquid bodies in static equilibrium. Since the free-body is in a state of static equilibrium, the algebraic sum of the components of external forces along the vertical axis must equal zero. Thus, remembering that the surface forces acting on the curved cylindrical surface have no components in the vertical direction, we obtain

$$\delta F \cos\theta = \delta W = \gamma\,\delta A' h \tag{2.7}$$

Now, as before, the pressure at a point P, located at a depth h from the surface, is given by the following limit:

$$p \equiv \lim_{\delta A \to 0} \frac{\delta F}{\delta A} = \lim_{\delta A \to 0} \frac{\gamma h(\delta A'/\cos\theta)}{\delta A'/\cos\theta} = \gamma h \tag{2.8}$$

Since θ is an arbitrary angle of inclination of the elemental area δA, the preceding result shows that the pressure remains the same at point P, irrespective of the orientation of the plane. In other words, *the liquid in static equilibrium exerts equal pressures in all directions at a given point, and the magnitude of pressure depends on the depth of the point from the free surface.*

2.1.2.2 *Static pressure at a point not directly under the free surface*

Let us now consider the problem of static pressure at a point P, which does not lie directly under the free surface, as shown in Figure 2.4. In this case, what is the significance of depth h in the equation for pressure, $p = \gamma h$? To answer this question, we proceed as follows. Let there be two points P and Q, lying in the same horizontal plane, such that the point Q is located directly under the free surface. Let us further consider a cylindrical free body with a constant elemental cross-sectional area δA, created by a

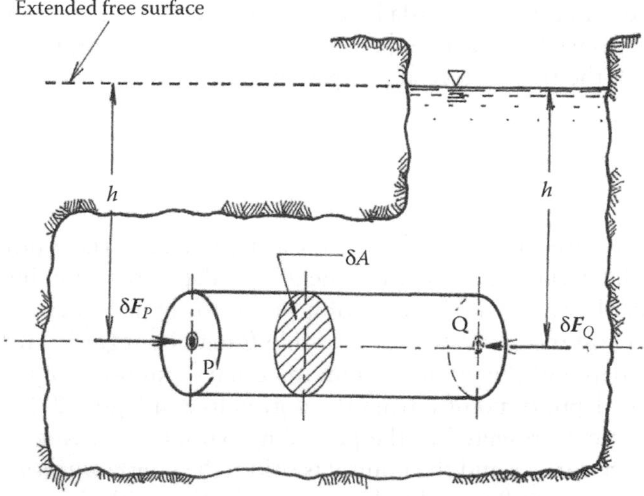

Figure 2.4 Pressure at a point P not directly beneath the free surface.

horizontal straight line generator and having two orthogonal end surfaces passing through points P and Q, as shown in Figure 2.4. Let there be resultant forces $\delta \mathbf{F}_P$ and $\delta \mathbf{F}_Q$ on the two elemental areas, passing through points P and Q, respectively. The forces acting on the curved cylindrical surface are oriented at right angle to the longitudinal (horizontal) axis of the cylinder; thus, these forces do not have components acting along the longitudinal axis of the free body. Since the body force δW (weight of the free body) acts vertically downward, it also has no component in the horizontal direction, including the axis of the free body. Since the free body is in a state of static equilibrium, the algebraic sum of the (scalar) components of external forces along the longitudinal axis must be equal to zero. Thus,

$$\delta F_P = \delta F_Q \tag{2.9}$$

(The reader is reminded that δF_P represents the magnitude of the vector $\delta \mathbf{F}_P$.) Now, by definition, the pressure at points P and Q are, respectively, given by the following limiting ratios:

$$p_P = \lim_{\delta A \to 0} \frac{\delta F_P}{\delta A} \tag{2.10a}$$

$$p_Q = \lim_{\delta A \to 0} \frac{\delta F_Q}{\delta A} \tag{2.10b}$$

Replacing δF_P in Equation 2.10a by δF_Q, using Equation 2.9, shows that the pressure at point P is exactly equal to the pressure at point Q. Since point Q is beneath the free surface at a depth h, the following transitive equality is true:

$$p_P = p_Q = \gamma h \tag{2.11}$$

Thus, in the case of a point lying beneath the free surface (for instance point Q), the variable h represents the actual depth of the point from the free surface. However, in the case of a point not lying under the real free surface (for instance point P), the variable h may be regarded to represent the virtual depth of the point from an imaginary extension of the free surface (see the depth of point P from the dotted line in Figure 2.4).

The argument presented in the preceding paragraphs assumes the existence of a free surface and the point P is said to be *communicating* with the free surface. A point P is said to be communicating with the free surface if the point P can be joined with another point S on the free surface through a curve, without leaving the liquid body at all.

2.1.3 Bernoulli's theorem

In this section, we intend to develop Bernoulli's theorem for an incompressible, ideal fluid (no friction) in a state of steady motion, based solely on energy consideration. According to the fundamental law of thermodynamics, energy can be neither created nor destroyed. Furthermore, all forms of energy are equivalent and their dimensions are the same as that of work—distance times force. In this section, we are particularly interested in two forms of energies: the *kinetic energy* and the *potential energy*. Briefly stated, the kinetic energy of a given mass is due to the velocity of the mass; while the potential energy is due to the location of the center of mass. The sum of *kinetic* and *potential energy* is also referred to as the *mechanical energy*. These forms of energy are further described, in some detail, in the subsequent sections.

2.1.3.1 Kinetic energy

The original enunciation of Newton's laws of motion relates to the dynamics of a point-mass, which assumes a vanishingly small volume for a given mass m. Also, while analyzing the rectilinear motion of a point-mass, the vector quantities can be easily treated as scalars, with a positive or a negative sign attached to them, to indicate the sense of direction. The advantage of using vector algebra over the ordinary scalar algebra in this case disappears.

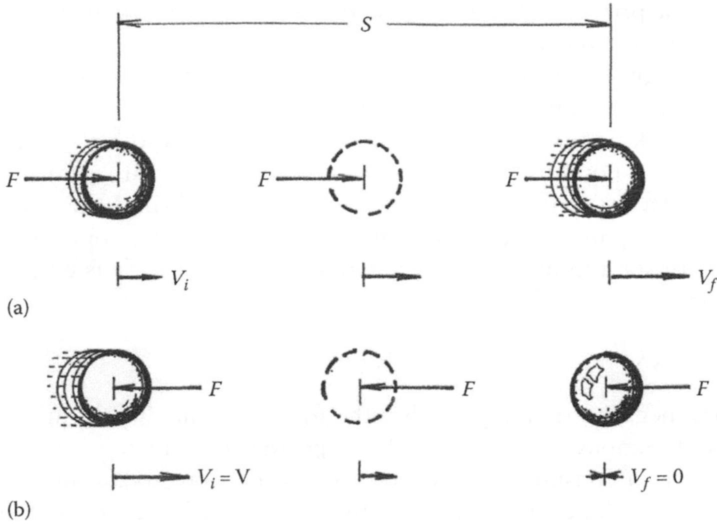

Figure 2.5 (a) Rectilinear motion of point-mass *m*, acted upon by an accelerating force *F*; and (b) rectilinear motion of point-mass *m*, acted upon by a retarding force *F*.

Let us suppose a point-mass m is initially moving in a straight line with a constant velocity (or speed) V_i. Let us also assume that a constant net force, F, acts on the system over a distance S, to impart a final velocity V_f, as shown in Figure 2.5a. The work done, W, by the force, F, on the point-mass, m, during displacement S, is given by the following equation:

$$W = F \times S \tag{2.12}$$

Since the net force acting on the point-mass is constant, the resulting acceleration of the mass should also be constant. From our knowledge of *kinematics* of rectilinear motion with a constant acceleration, the following formula is applicable:

$$\frac{\left(V_f^2 - V_i^2\right)}{2a} = S \tag{2.13}$$

where $a = F/m$, the constant acceleration of the point-mass m. Combining the last two equations yields

$$W = \frac{mV_f^2}{2} - \frac{mV_i^2}{2} \tag{2.14}$$

Based on the principle of conservation of energies, the work done by a force on a system of mass m must cause an increase in the energy of the system. Thus, the right-hand side of the preceding equation should represent the increase in the energy of the system m.

Let us look at the physics of this problem from another perspective. Let there be a point-mass m, translating with a uniform velocity V; and let there be a net uniform opposing force F, which brings the mass m to a complete stop over a distance S (Figure 2.5b). Since the opposing force remains constant in magnitude, the work done in this case is given by the following:

$$W = -F \times S \tag{2.15}$$

where the negative sign implies that the force and the displacement are in opposite directions. Again, since the opposing force remains constant, it must impart a constant acceleration (or retardation) to the mass. Thus, in this case the kinematic relationships for the rectilinear translation of a point-mass with constant acceleration must hold; in particular,

$$S = \frac{V_f^2}{2a} - \frac{V_i^2}{2a} = -\frac{V^2}{2a} \tag{2.16}$$

because the final velocity $V_f = 0$ and the initial velocity is V. Now, combining Equation 2.16 with Equation 2.15 yields the following:

$$W = \frac{mV^2}{2} \tag{2.17}$$

It is interesting to note that the right-hand side of the previous equation contains the variable that solely describes the kinetic condition of the mass m, prior to the application of the opposing force. In other words, the quantity, $mV^2/2$, is indeed an attribute of the moving material object, oblivious to the future encounter with the force. And, also, this quantity is exactly equal to the work done on the point-mass in bringing it to a complete rest. It is therefore understandable that this quantity is designated in the literature as the *kinetic energy* of the point-mass translating with velocity V. The adjective *kinetic* refers to the fact that this quantity, for a given mass m, depends only on its velocity; and *energy* refers to the fact that its dimension is the same as that of work; it also represents the capacity of the moving object to do work on the surrounding until it comes to a complete rest.

The development given hitherto—in particular the one leading to Equation 2.14—is an affirmation of the fact that the general work-energy equation is true in the case of a rectilinear motion of point-mass subject to a constant net force. The work-energy equation is, however, more general in scope. It is valid for any curvilinear trajectory of the point-mass, acted upon by a variable (as opposed to a constant) force. It can be obtained by integrating Newton's second law of motion along an actual trajectory of the point-mass. Instead, as stated previously, we intend to pursue more physical and geometrical intuition in the derivation of this equation. It is, therefore, desirable to obtain the general work-energy equation based on heuristic arguments, starting with Equation 2.14. To achieve this, we proceed as follows.

With regard to Equation 2.14, it is worth stressing that the actual magnitude of either the constant force, F, or the distance S, does not appear, per se, in this equation, as long as the work, W, truly represents the work done by the impressed force on the point-mass. If the impressed force is variable, the work done by the variable force $\boldsymbol{F}(\boldsymbol{r})$ on the point-mass during its travel from point $\boldsymbol{r}_1$ to $\boldsymbol{r}_2$ can be obtained from the following integral (Figure 2.6):

$$W = \int_{r_1}^{r_2} \boldsymbol{F} \cdot d\boldsymbol{r} \tag{2.18}$$

In the evaluation of work, it is understood that the integration in the preceding equation is carried out along the actual trajectory from position $\mathbf{r}_1$ (point 1, in Figure 2.6) to position $\mathbf{r}_2$ (point 2, in Figure 2.6). This work on the point-mass must cause an increase in the kinetic energy of the point-mass (Equation 2.14). Thus,

$$W = \int_{r_1}^{r_2} \boldsymbol{F} \cdot d\boldsymbol{r} = \frac{mV_2^2}{2} - \frac{mV_1^2}{2} \tag{2.19}$$

where V_1 and V_2 are the speeds of the point-mass at points 1 and 2, respectively.

2.1.3.2 Potentials and the potential energy

The treatment by Housner and Hudson (1959) on this subject is straightforward and to the point. The discussion here follows, in parts, their development rather closely. It is evident from Equation 2.19 that while the right-hand side depends on the conditions at the two end points, point 1 and point 2, only, the work integral depends on the actual trajectory or the path. At this juncture, we wish to ask the following question: Under what

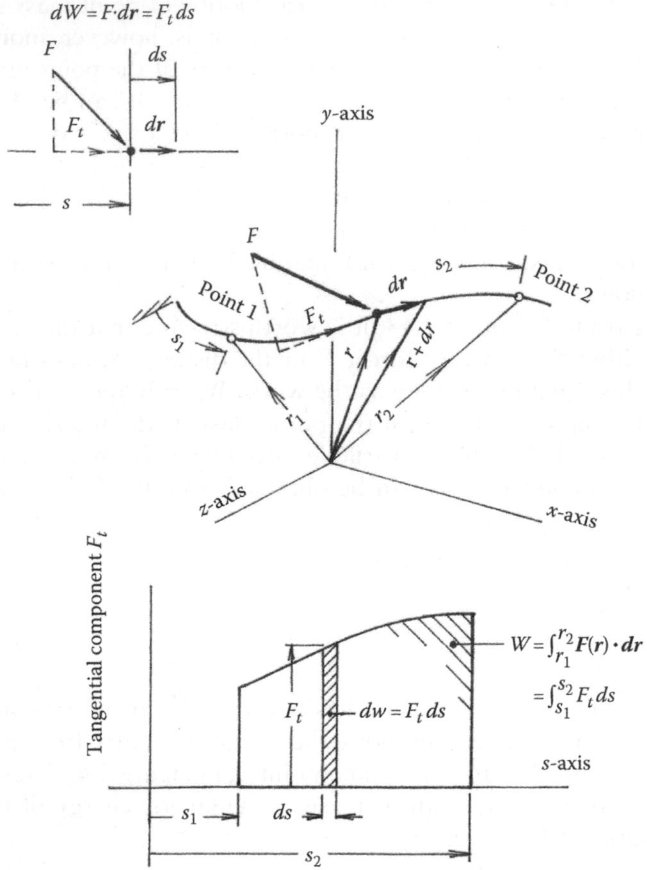

Figure 2.6 Definition sketch for work-energy equation.

conditions the work integral is independent of the actual path and depends only on the lower and upper limits of integration? The answer is deceptively simple: Whenever the infinitesimal work $\boldsymbol{F}{\cdot}\boldsymbol{dr}$ represents an *exact differential* of a differentiable function $\Phi = \Phi(x, y, z)$, the line integral is independent of the actual path and depends only on the lower and the upper limits of integration, because

$$W = \int_{r_1}^{r_2} \boldsymbol{F} \cdot \boldsymbol{dr} = \int_{\Phi_1}^{\Phi_2} d\Phi = \Phi_2 - \Phi_1 \tag{2.20}$$

When the work integral is independent of the actual path, the force is said to be derivable from a scalar function $\Phi(x, y, z)$: for

$$F \cdot dr = F_x\,dx + F_y\,dy + F_z dz \tag{2.21a}$$

$$d\Phi = \frac{\partial \Phi}{\partial x}dx + \frac{\partial \Phi}{\partial y}dy + \frac{\partial \Phi}{\partial z}dz \tag{2.21b}$$

and the following equalities hold

$$F_x = \frac{\partial \Phi}{\partial x}; \quad F_y = \frac{\partial \Phi}{\partial y}; \quad F_z = \frac{\partial \Phi}{\partial z}$$

because the differentials dx, dy, and dz are arbitrary. The scalar function Φ is also called the *potential function*.

It is partly due to tradition and partly due to convenience that we define another potential function U that is negatively equal to Φ so that

$$\int_{U_1}^{U_2} dU = -\int_{\Phi_1}^{\Phi_2} d\Phi = -\int_{r_1}^{r_2} F \cdot dr = -\left(\frac{mV_2^2}{2} - \frac{mV_1^2}{2}\right) \tag{2.22}$$

The function U is called the *potential energy associated with the force field*, and it is *meaningfully* defined only when the work integral is independent of the path. With the introduction of new potential $U = -\Phi$, the force can be derived as follows:

$$F_x = -\frac{\partial U}{\partial x}; \quad F_y = -\frac{\partial U}{\partial y}; \quad F_z = -\frac{\partial U}{\partial z}$$

or

$$F = -\left(i\frac{\partial U}{\partial x} + j\frac{\partial U}{\partial y} + k\frac{\partial U}{\partial z}\right) = -\nabla U$$

where the operator ∇ (pronounced as *del*) $\equiv \left(i\frac{\partial}{\partial x} + j\frac{\partial}{\partial y} + k\frac{\partial}{\partial z}\right)$ operates on any differentiable scalar function, such as $U = U(x, y, z)$, and yields, as a result of this operation, the gradient of the scalar function. Thus, the force vector $F(x, y, z)$ equals the negative gradient of the potential function U.

The role of tradition in introducing the function U lies in the fact that we like to see the force directed from a higher value to the lower value of the potential U (just as we like to perceive that the heat flows in the direction of decreasing temperature). The convenience of introducing U for $-\Phi$ lies in the fact that the *principle of conservation of mechanical energy* can now be stated more succinctly:

$$U_2 - U_1 = -\left(\frac{mV_2^2}{2} - \frac{mV_1^2}{2}\right)$$

or

$$U_1 + \frac{mV_1^2}{2} = U_2 + \frac{mV_2^2}{2} = \text{a constant} \tag{2.23}$$

which asserts that the *sum* of potential and kinetic energies (and not the difference between Φ and the kinetic energy) remains constant, in a force field where the work integral is independent of the path. Such a force field is, therefore, said to be a *conservative force field*: because, *the sum of the potential energy and the kinetic energy is preserved (or conserved)*. Thus, only in a conservative force field, *the principle of conservation of mechanical energy holds*.

The principle of conservation of mechanical energy is more general than required for the study of groundwater. For this study, we want to know whether the gravitational force field near the surface of the earth represents a conservative force field or not! To ascertain the nature of the gravitational force field near the earth surface, the following illustration is relevant.

Let us assume that a point-mass m moves along a certain path from position 1 to position 2 in a gravitational field near the surface of the earth, as shown in Figure 2.7. Throughout its motion, the point-mass is acted upon by a constant force due to its weight, that is, $\boldsymbol{F} = m\boldsymbol{g}$. During a small (infinitesimal) displacement $d\boldsymbol{r}$, the work done by this force is

$$\begin{aligned} \boldsymbol{F} \cdot d\boldsymbol{r} &= m\boldsymbol{g} \cdot d\boldsymbol{r} = mg \times dr \times \cos(\pi - \beta) \\ &= -mg \times dr \times \cos\beta = -mg \times dy \end{aligned} \tag{2.24}$$

where β denotes the direction angle (see Figure 2.7). It follows from the last equality that

$$\boldsymbol{F} \cdot d\boldsymbol{r} = d(-mgy)$$

which represents an *exact differential*. Thus, the work integral is independent of the path and the gravitational force field near the earth surface constitutes a conservative force field. In this case, it is, therefore, meaningful to define the differential of the potential energy as

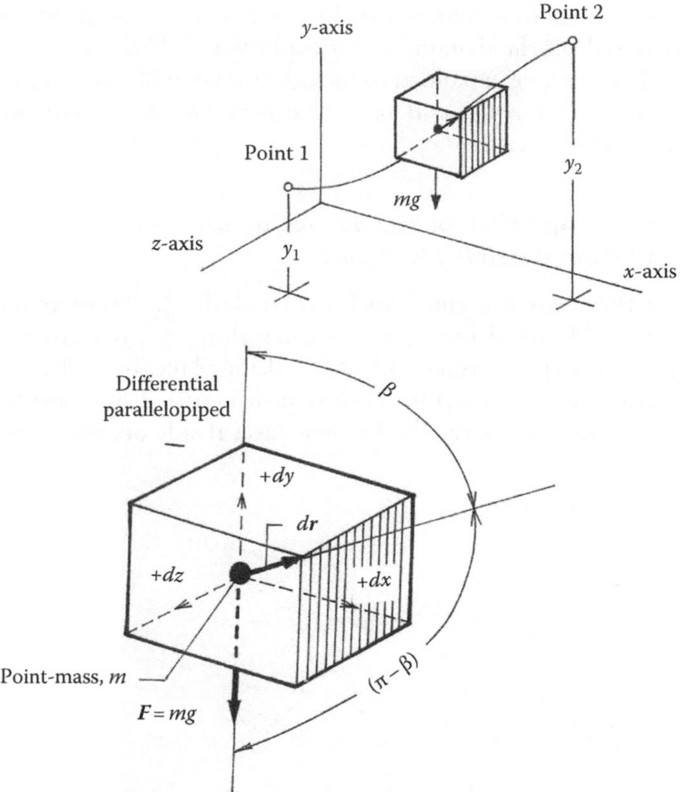

Figure 2.7 Force field due to gravity near the surface of the earth.

$$dU = -\boldsymbol{F} \cdot \boldsymbol{dr} = mg\,dy \tag{2.25}$$

which, on integration, yields

$$U = (mg)y + C$$

The constant of integration, C, can be made to vanish by arbitrarily choosing $U=0$ at $y=0$. Since the origin of the coordinate system is arbitrary, this demonstrates that the potential energy can be set equal to zero at any arbitrary (but convenient) level. Thus, the increase in the potential energy as the point-mass moves from point 1 to point 2 (Figure 2.7) is given by the following:

$$U_2 - U_1 = mg(y_2 - y_1) \tag{2.26}$$

Finally, a few concluding remarks are in order. The developments presented so far are based on the dynamics of a point-mass. The notion of a point-mass may, however, be expanded to include a system of mass, m, with non-negligible size, *if the rotational velocities are negligible in comparison to the translational velocities.*

2.1.3.3 Energy equation for steady motion of an incompressible ideal fluid

We extend the ideas presented so far to include the laminar motion of an incompressible ideal fluid (nonviscous) along a streamline. For this purpose, we like to perceive a steady flow taking place through an infinitely small diameter of a streamtube as shown in Figure 2.8. A streamtube is an imaginary tube inside the fluid whose lateral side (as opposed to cross

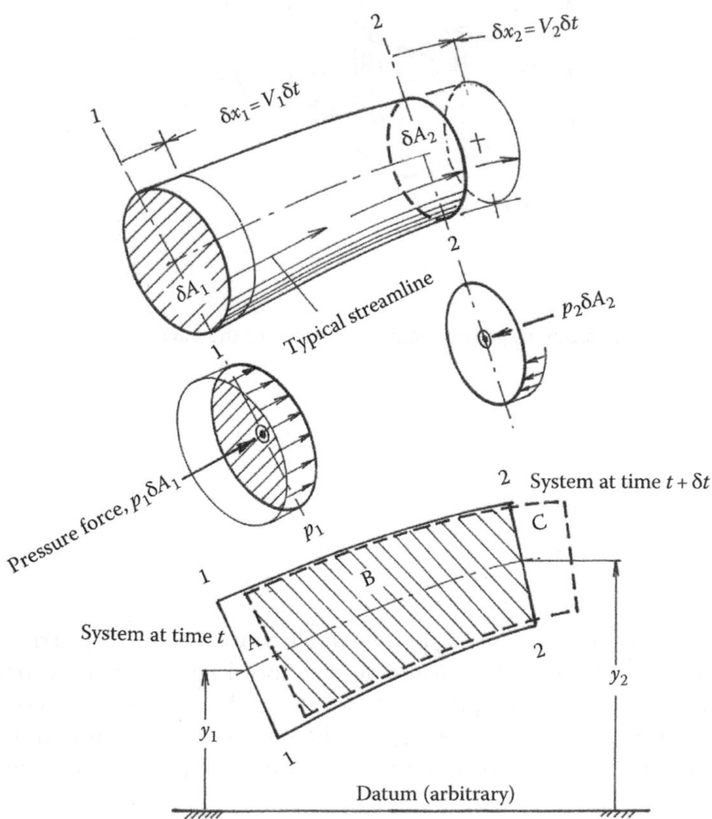

Figure 2.8 Definition sketch for Bernoulli's equation.

sections) is generated by streamlines. Thus, no flow can either enter or leave the streamtube, except through the cross-sectional areas.

In fluid mechanics, we use the term *system* with a special connotation: It refers to a portion of fluid bound by an enclosing surface such that the system always refers to the same matter, irrespective of the configuration of the enclosing surface. Now, let a system of mass m occupy a volume of the streamtube at any instant of time t, as shown by the solid lines in Figure 2.8. Let the same system of mass m occupy another volume, shown by dotted lines in the figure, at some subsequent time $t+\delta t$, where δt represents an infinitesimal increment of time t. Thus, in conformity with the principle of conservation of energy, the work done by the external forces on the system of mass m, during time interval δt, must equal the increase of the mechanical energy of the system m during the same time interval δt. In other words, the work done by external forces, during the time interval δt,

$$\delta W=\delta(KE)+\delta(U) \tag{2.27}$$

where the first and the second terms on the right-hand side represent the increase in the kinetic and the potential energy of the system m, respectively.

To evaluate the work done by external forces, as time lapses from t to $t+\delta t$, we consider the pressure forces acting on the two cross sections, designated as 1-1 and 2-2 in Figure 2.8. Thus,

$$\delta W=\delta W_1+\delta W_2 \tag{2.28}$$

where δW_1 and δW_2 represent the work done by pressure forces on the two cross sections 1-1 and 2-2, respectively, as these cross sections advance in the direction of flow during the time interval δt. Furthermore (see Figure 2.8),

$$\delta W_1 = \left(p_1 \delta A_1\right) V_1 \delta t \tag{2.29a}$$

$$\delta W_2 = -\left(p_2 \delta A_2\right) V_2 \delta t \tag{2.29b}$$

where

the subscripts, 1 and 2, refer to the lower and upper cross sections

the variables p, δA, and V refer to pressure, cross-sectional area, and the speed of flow, respectively

The negative sign in Equation 2.29b implies that the pressure force and displacement are opposite to each other on cross section 2-2. Since the fluid is incompressible and the motion is steady, it follows that

$$\delta A_1 V_1 \delta t = \delta A_2 V_2 \delta t = \delta(\mathrm{Vol}) \tag{2.30}$$

where δ(Vol) represents the volume of fluid crossing any cross section during time interval δt. Thus,

$$\delta W = (p_1 - p_2)\delta(\text{Vol}) \tag{2.31}$$

Now, to find the change in the mechanical energy of the system of mass m as the time lapses from t to $t+\delta t$, we argue as follows: By definition, this change in mechanical energy equals the difference between the mechanical energy of the system of mass, m, at time $t+\delta t$ and the mechanical energy of the same system of mass at time t. As time lapses, the (spatial) configuration of the system of mass m changes. The new configuration of the system of mass m is characterized by the fact that space A has been subtracted and space C has been added to the previous configuration of the same system of mass m (see Figure 2.8). In other words, the change in the mechanical energy of the system m is equal to the difference between the mechanical energies of mass occupying space C and that occupying space A, because the mass occupying the common space B does not contribute to the change due to the assumption of a steady motion. Thus,

$$\delta(KE) = \rho\delta(\text{Vol})\frac{V_2^2}{2} - \rho\delta(\text{Vol})\frac{V_1^2}{2} \tag{2.32a}$$

$$\delta(U) = \rho g\,\delta(\text{Vol})y_2 - \rho g\,\delta(\text{Vol})y_1 \tag{2.32b}$$

Now combining Equation 2.27 with Equations 2.31 and 2.32a,b and dividing throughout by the factor ρg δ(Vol) as $\delta t \to 0$, the following equation is obtained:

$$\frac{(p_1 - p_2)}{\rho g} = \frac{V_2^2}{2g} - \frac{V_1^2}{2g} + (y_2 - y_1) \tag{2.33a}$$

or

$$\frac{p_1}{\gamma} + y_1 + \frac{V_1^2}{2g} = \frac{p_2}{\gamma} + y_2 + \frac{V_2^2}{2g} = \text{a constant} \tag{2.33b}$$

where γ denotes the specific weight of the fluid. Equation 2.33b represents one form of the Bernoulli equation, after the Swiss physicist and mathematician Daniel Bernoulli (1700–1782), who published it in 1738 along with his famous work, *Hydrodynamica*.

Each term in the Bernoulli equation, as stated earlier, represents the energy per weight. In US customary units, these terms are expressed in foot (ft·lb/lb = ft), while in SI units, meter (m·N/N = m) is frequently used.

Since each term in Equation 2.33b has the dimension of length [L], the first three terms, starting from the left, are designated as the *pressure head*, the *elevation head*, and the *velocity head*, respectively. Their sum represents the *total energy head*. Thus, according to the Bernoulli equation, the total energy head remains the same along a streamline, as long as the fluid is ideal (nonviscous), incompressible, and the flow is steady. As a further consequence, along a horizontal streamline ($y_1 = y_2$), the velocity increases at a point where pressure decreases. Finally, when velocity heads are zero, the preceding equation degenerates to an unreal situation, where, in the absence of velocity, the existence of a streamline, or a streamtube, becomes imaginary. Despite this degeneration, the Bernoulli equation does conform to the pressure distribution in stagnant fluids, for it reduces to

$$p_1 - p_2 = \gamma(y_2 - y_1) \tag{2.34}$$

when velocity tends to zero.

2.2 PRELIMINARIES FROM SOIL MECHANICS

2.2.1 Porosity of soil or porous medium

The *nominal volume (or the total volume)* represented by a sample of soil or porous medium can be divided into two parts: one occupied by the solid skeleton (or the granular matrix) of soil, and the other by the void spaces in the soil. The sum of these two parts completely comprises the nominal volume of the sample. For such a soil sample, the porosity can be defined as the ratio of volume of void space to the nominal volume of the sample (Figure 2.9a). In mathematical notation,

$$n = \frac{V_v}{V} \tag{2.35}$$

where

n is the (volumetric) porosity, a dimensionless quantity [L^3/L^3]
V_v is the volume occupied by void spaces [L^3]
V is the nominal (or total) volume of sample [L^3]

Porosity is sometimes represented as a percentile ratio. Since soil structure is highly random on a microscopic scale, it is understood that the nominal volume is *just* large enough so that the ratio defined in Equation 2.35 represents a stable average value.

Similar to the notion of volumetric porosity, we have the notion of areal porosity. For this purpose, we assume that a typical cross section of the

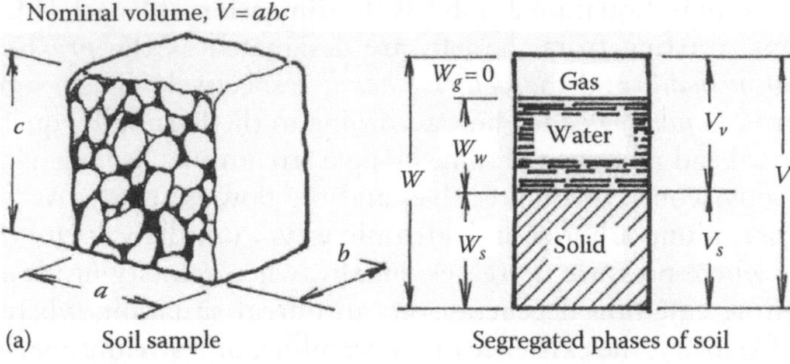

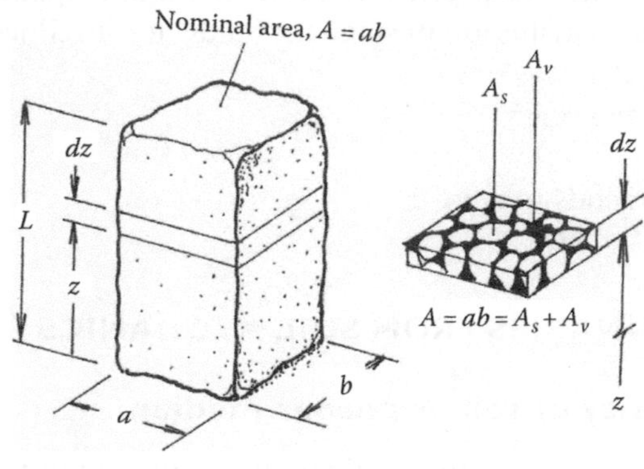

Figure 2.9 Definition sketches: (a) volumetric porosity and (b) areal porosity.

soil sample consists of two parts: one representing the area occupied by the grains, and the other representing the area occupied by void spaces (Figure 2.9b). The sum of these areas comprises the nominal cross-sectional area of the soil sample. We can now define

$$m = \frac{A_v}{A} \tag{2.36}$$

where

m is the areal porosity, a dimensionless quantity $[L^2/L^2]$
A_v is the area of void space $[L^2]$
A is the nominal cross-sectional area of the soil $[L^2]$

It can be shown that the average value of areal porosity is the same as the value of volumetric porosity; for

$$\bar{m} = \frac{1}{L}\int_0^L m(z)dz = \frac{1}{AL}\int_0^L Am(z)dz = \frac{1}{V}\int_0^L A_v(z)dz = \frac{V_v}{V} = n \tag{2.37}$$

The first equality follows from the mathematical definition of an average value, $\bar{m}$, and the last equality follows from the definition equation (2.35). Henceforth, we shall simply use the term *porosity* to indicate the areal or the volumetric porosity of the soil sample.

Although porosity is a good indication of the storage capacity of the soil or a porous medium to store groundwater in its pore spaces, this physical property loses much of its significance during the consolidation of soils. During consolidation, both the numerator and the denominator in definition equation (2.35) undergo changes. For such situations, another physical property, called *void ratio*, becomes more pertinent. It is defined in the following section.

2.2.2 Void ratio, water content, and degree of saturation

The void ratio is defined by the following equation:

$$e = \frac{V_v}{V_s} \tag{2.38}$$

where

e is the void ratio, a dimensionless quantity [L^3/L^3]
V_s is the volume of soil skeleton [L^3]

In the preceding definition, it is understood that V_s is incompressible compared to V_v.

Another physical property especially for fine-grained soils is called the *water content*, w. It is defined as

$$w(\%) = 100\frac{W_w}{W_s} \tag{2.39}$$

where

W_w is the weight of water in a soil sample [ML/T^2]
W_s is the weight of soil skeleton in a soil sample [ML/T^2]

The weight of soil skeleton, W_s, also represents the oven-dry weight of the soil sample at a standardized temperature range of 105°C–115°C. Similar to void ratio, in the definition of water content, the weight of water is referred to the unchanging weight of soil solids. Sometimes, especially in the literature on hydrogeology, volumetric ratio is used to express moisture content. The volumetric water content θ is defined as the ratio of contained volume of water V_w to the nominal volume of the sample V, that is, $\theta = V_w/V$.

The void spaces in a soil, especially above water table, may not be fully saturated. In other words, a part of the void space may be filled with air. For such situations, the degree of saturation is defined as

$$S(\%) = 100\frac{V_w}{V_v} \tag{2.40}$$

where

S is the degree of saturation in percent

V_w is the volume of water in the soil sample [L^3]

2.2.3 Total pressure, porewater pressure, and effective pressure

The unit weight (or specific weight) of a substance represents the ratio of the weight to the confining volume of the substance. In the case of natural soils, the weight of the sample varies depending on its water content. It is therefore meaningful to define at least two kinds of unit weights: the dry unit weight, γ_d, for oven-dry soils and the saturated unit weight, γ_{sat}, for the saturated soils.

Now, let us consider a submerged sample of soil held in a container as shown in Figure 2.10. We shall assume that the submerged soil sample as well as the inundating water is in a state of static equilibrium. We shall also assume the submerged soil behaves like a fluid; in particular, we shall

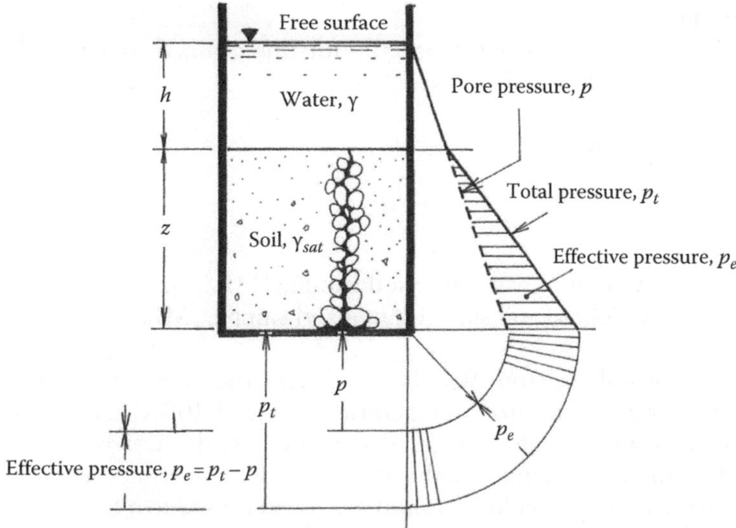

Figure 2.10 Definition sketch for total, porewater, and effective pressure.

assume no shear stress between the soil and the interior surface of the container. Under such stipulations, the total pressure, p_t, at a point on the bottom of the container is given as

$$p_t = \gamma h + \gamma_{sat} z \tag{2.41}$$

In the preceding equation, γ denotes the unit weight of water and the depths h and z are shown in the figure. The hydrostatic pressure inside a pore (porewater) in contact with the bottom is given by

$$p = \gamma(h + z) \tag{2.42}$$

as long as the pore space is *communicating* with a point on the free surface of water. The effective pressure is defined as the difference between the total and the hydrostatic pressure. Thus, the effective pressure is given as

$$p_e \equiv p_t - p = \left(\gamma_{sat} - \gamma\right) z \tag{2.43}$$

In the preceding equation, $(\gamma_{sat} - \gamma)$ represents the so-called submerged unit weight of the soil. It is interesting to note that the effective pressure does not depend on the depth of submergence, h. In soil mechanics, the porewater pressure is also called the neutral stress and it does not affect directly either the shear resistance or the compressibility of soils. It is the effective pressure (or, more precisely, the effective stress) that plays a paramount role in the strength and deformation mechanics of soils. This assertion is commonly known as the *principle of effective stress* in the literature on soil mechanics and geotechnical engineering (Terzaghi, 1943; Lambe and Whitman, 1969).

2.3 CONTINUUM CONCEPT OF A POROUS MEDIUM

The idea of a continuum is as old as the analytical thinking: Is the essence of physical universe discrete or continuous? Both notions have been equally present in our abstract thoughts: sometimes complementing and sometimes contradicting each other! If the history of scientific thoughts is any indication, it appears that our analytical effort has been more on the part of continuous—as manifestly borne out by the enormous development of *infinitesimal* calculus and its predominant role in the scientific and engineering curricula. Part of this predominance is due to the remarkable success of analytical methods based on infinitesimal calculus in formulating and finding acceptable answers to a vast array of complex engineering problems. Nevertheless, all mathematical entities are a figment of imagination. Their usefulness or lack of it in solving a particular physical problem can only be determined by empiricism. Rationalism alone cannot pass judgment on its own effectiveness.

In the past, be it the analysis of solids, fluids, or gases (with the possible exception of rarified gases), we have used the notion of continuum quite effectively, despite the fact that all matter is discrete on a sufficiently small (*microscopic*) scale. Can we use the same notion of a continuum, coupled with the mathematical formulation based on partial differential equations utilizing the limiting process of infinitesimal calculus, in the analysis of flow through a porous medium, where the discrete nature of the granular material is visible even to unaided eyes? The short answer is yes: provided, the smallest dimension of the flow region of practical interest is sufficiently large (by many orders of magnitude) in comparison with the largest intergranular distance between particles of the porous medium. From a purist's perspective, the issue is the same whether we analyze a fluid continuum or flow through a porous medium. It is the relative size of the region of practical interest, in comparison to the intergranular distance, that matters.

In the foregoing, we have left intentionally the issue in its bare essentials. A more involved discussion only further complicates it. If we agree to use the mathematical formulation based on partial differential equations, involving limiting processes of infinitesimal calculus, the issue of finding or extrapolating the values of the *field variables* (e.g., pressure, porosity, density, coefficient of permeability) down to the limit of a mathematical point deserves a more careful attention. This aspect of the problem is further discussed in the subsequent sections.

2.3.1 Notion of porosity in a porous medium

As a point of departure, we shall start a more careful discussion on the concept of porosity in this section. For this purpose, we refer to the illustration in Figure 2.11a, which represents a sequence of concentric spheres centered at a point P with diminishing radii:

$$r_1 > r_2 > r_3 > \cdots > r_n \quad \text{where} \quad r_n \to 0 \text{ as } n \to \infty \tag{2.44}$$

We can also conceive a corresponding sequence of porosity whose general term is computed as

$$n_n = \frac{(V_v)_n}{(V)_n}; \quad n = 1,2,3,\ldots,\infty \tag{2.45}$$

where

n_n represents the porosity based on a spherical *nominal volume*, $(V)_n = 4/3\,\pi r_n^3$, of the soil sample centered at P with a radius r_n

$(V_v)_n$ represents the total void space contained in this sample

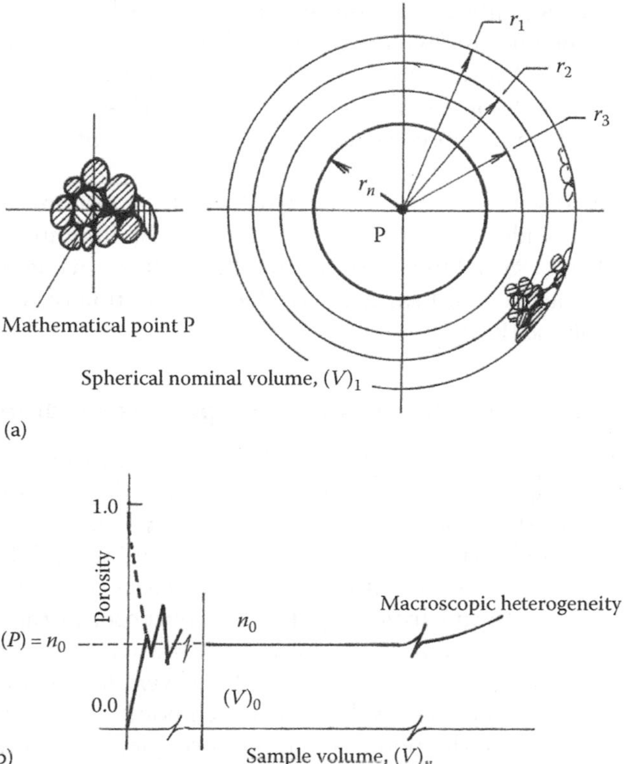

Figure 2.11 (a) A sequence of concentric spheres, and (b) a schematic plot of porosity versus spherical sample size.

In Figure 2.11b, the value of porosity versus the volume of spherical soil sample, $(V)_n$, is schematically plotted. It can be perceived that the porosity values will indicate rapid fluctuations as the radius of sample falls below a value r_0, representing a certain order of magnitude of intergranular dimension. It is also perceivable that the porosity value will eventually converge either to 0 or 1, depending upon the location of point *P*, whether it is occupied by the grain or the pore space. As the nominal size of the sample increases beyond r_0, or $(V)_0 = 4/3\pi r_0^3$, the porosity values will stabilize because of the presence of a large (statistically speaking) number of grains. If the sample size is further increased, the effect of *macroscopic heterogeneity* of the porous medium may creep in.

The foregoing discussion about the behavior of porosity is schematically shown in Figure 2.11b. If the medium is *locally homogeneous on macroscopic scale*, the porosity values n_n will remain almost constant as the sample radii r_n increase in the neighborhood of r_0. In other words, porosity $n_n = n_0$ does not depend on the value of r_n, as long as $r_0 < r_n < r_u$. If we disregard the microscopic fluctuations, the constant value of porosity, n_0, may be regarded as the extrapolated value of porosity as the sample size r_n tends to zero without leaving the point P. Thus, this provides a rational procedure for defining porosity at a mathematical point P of the porous continuum.

For further discussion on the effectiveness of continuum approach in mechanics, the reader is referred to the literature by Prandtl and Tietjens (1934), Hubbert (1957), Hodge (1970), and Bear (1972). In connection with the flow in porous media, Bear has introduced the notion of representative elementary volume (REV).

2.3.2 Notion of rectilinear flow in a porous medium

The movement of groundwater through a porous medium is generally sluggish and complex. The actual path taken by a fluid particle (parcel) through the interstices of the porous medium on a microscopic level is rather irregular and tortuous in nature, as shown in Figure 2.12a. In such a medium, the concept of a rectilinear motion ceases to exist, and it can only be defined in an idealistic way. For this purpose, in the theoretical analysis of groundwater flow, we shall replace the actual porous medium by a *fictitious homogeneous medium*, in which every field variable associated with the porous medium is defined at each point (x, y, z), irrespective of the fact of whether the point is actually occupied by a solid grain or a void space. We shall also regard the groundwater as an *ideal continuum*, which permeates throughout the *fictitious porous medium*, and the

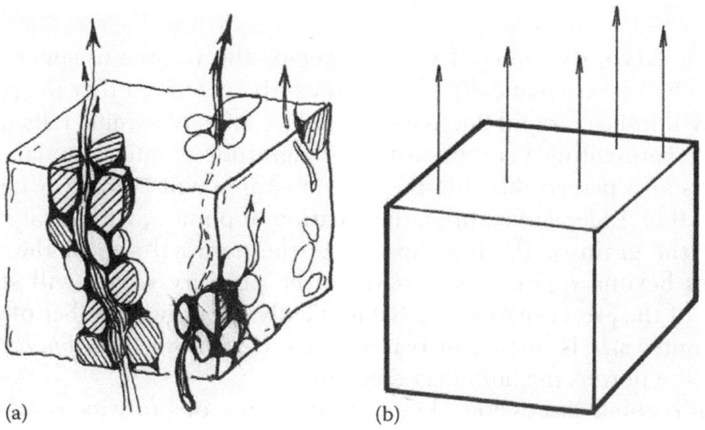

Figure 2.12 (a) Tortuous flow through porous medium, and (b) equivalent rectilinear flow through porous medium.

relevant flow variables of this ideal continuum are defined at each *mathematical point* for all times. Under such idealized conditions, we shall designate the groundwater flow as a rectilinear flow field if the theoretical (as opposed to actual) velocity field remains constant (Figure 2.12b). The usefulness of such analyses, off course, rests on the anticipation that the theoretical flow should represent, on a macroscopic scale, an average flow field. In this regard, the analyst plays a paramount role in interpreting the theoretical results.

2.3.3 Specific discharge and seepage velocity

The discharge of a liquid through an orifice is defined as the volume of liquid crossing the orifice in a unit time. If the flow is uniform and the velocity vector V is normal to the area A of the orifice, as shown in Figure 2.13a, then discharge Q is given by the following equation:

$$Q = AV \tag{2.46a}$$

On the other hand, if the flow is uniform but the velocity vector acts at an oblique angle to area as shown in Figure 2.13b, the discharge is given by the following equation:

$$Q = AV\cos\theta \tag{2.46b}$$

where θ represents the angle between velocity vector V and unit vector e_n, acting normal to the orifice area (see Figure 2.13b). It is evident from

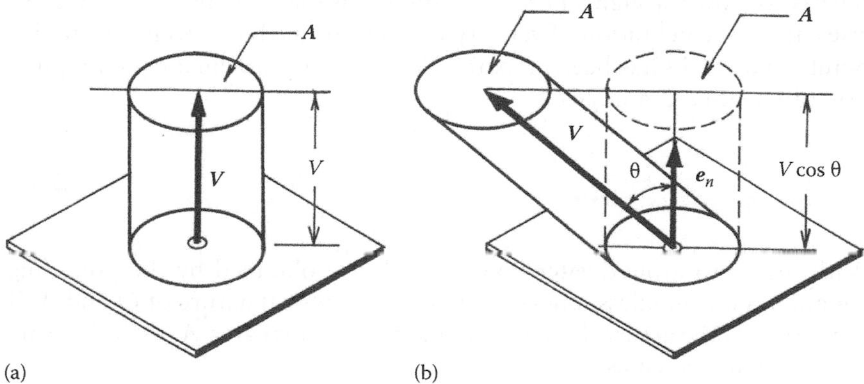

Figure 2.13 Definition sketch for discharge of liquid through an orifice: (a) velocity vector normal to the orifice area, and (b) velocity vector at an oblique angle to the orifice area.

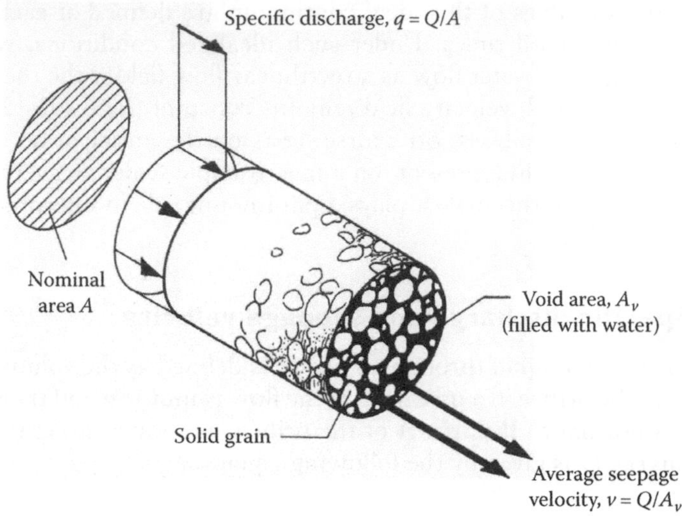

Figure 2.14 Illustration of relationship between specific discharge and average seepage velocity.

Equation 2.46b that the normal component of velocity vector is instrumental in transporting the fluid across the orifice. The tangential component of velocity, however large, is inconsequential in contributing to the discharge.

The preceding relationships are purely geometric in nature. However, an important fact in developing the preceding equations is the perception that the entire area of the orifice is available for flow. This may not be true in the case of flow through a porous medium. It is therefore necessary to exercise care in the definition of velocity in a porous medium. To illustrate this point, Figure 2.14 has been prepared. A just facile mathematical manipulation of Equation 2.46a yields

$$V = \frac{Q}{A} \tag{2.47}$$

With regard to groundwater flow, the velocity obtained by the preceding equation is meaningless unless we define the physical nature of Q and A. If Q represents the actual discharge across the nominal area A, then there are two different velocities:

$$q \equiv V = \frac{Q}{A} \tag{2.48a}$$

and

$$v = \frac{Q}{A_v} \tag{2.48b}$$

where

A represents the nominal (total) cross-sectional area
A_v is the total void area contained in A (Figure 2.14)

In the literature on groundwater, vector q is called the Darcy velocity and its magnitude q the *specific discharge*. Velocity v represents the *average seepage velocity*. If Q represents the actual discharge across the cross-sectional area A, then

$$Q = q\,A = v\,A_v \tag{2.49}$$

or,

$$q = v\frac{A_v}{A} = v\,n \tag{2.50}$$

Since porosity n is always less than one, it follows from the preceding equation that *the specific discharge q is always less than the average seepage velocity v.*

2.4 STAGNANT GROUNDWATER AND ZERO-GRADIENT OF PIEZOMETRIC HEAD

A sample of soil is shown completely submerged in Figure 2.15. We shall assume that the entire pore space is filled with water, which is in a state of hydrostatic equilibrium. The phrase *hydrostatic equilibrium* implies that the water body occupying the pore space of the granular medium is *stagnant, contiguous, and communicates* with the free water surface. If we insert a piezometric tube at a point P, the rise of water in the piezometer will be equivalent to the pore pressure at the point of insertion (Figure 2.15). In other words, if h denotes the rise of water in the piezometer, p the porewater pressure, and γ the specific weight of water, then $h = p/\gamma$. If z denotes the elevation of the point P, with respect to an arbitrary datum, the sum $z + h = z + p/\gamma$ is called the *piezometric head or piezometric level* at point P, and it is denoted by ϕ. In the case of a stagnant body of porewater, the piezometric head ϕ at all points represents the same constant elevation of the free water surface (see Figure 2.15). The last assertion is

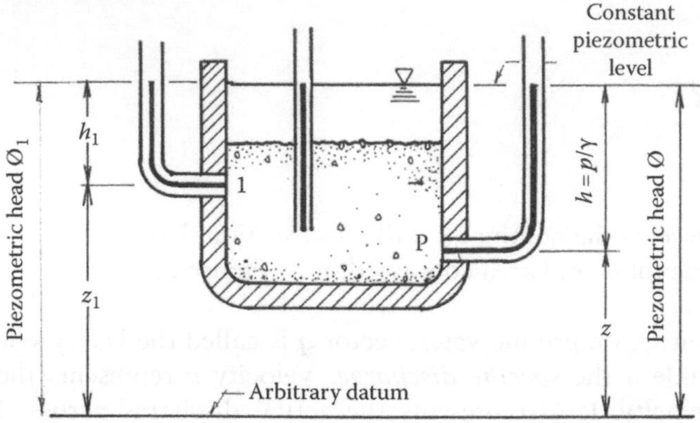

Figure 2.15 Hydrostatic water and gradient of piezometric head.

independent of the nature of the submerged porous medium with respect to its homogeneity and isotropy. Thus, we come to the conclusion that *in the case of stagnant groundwater the gradient of the piezometric head must vanish.*

It is not difficult to imagine that a nonzero Darcy velocity vector q at any point corresponds with a nonzero gradient vector, $\nabla\phi$, at the same point in the groundwater flow field. Since $q=0$ corresponds with $\nabla\phi=0$, it is obvious that, for a sufficiently small magnitude of q, the relationship between vector q and the gradient vector $\nabla\phi$ must be proportional (this follows from the fact that all functional relationships in a sufficiently small neighborhood are linear, provided the functions are—casually speaking—not pathological). In the present case, this proportional relationship is, however, between two vector quantities (as opposed to scalar quantities). Thus, the constant of proportionality could, in general, be a *constant matrix of proportionality*. If, however, the porous medium is isotropic, then the vectors q and $\nabla\phi$ are related by a *scalar* constant of proportionality. In other words,

$$q = k\nabla\phi \tag{2.51}$$

where k is the scalar constant of proportionality. It is apparent from Equation 2.51 that the specific discharge vector q and the gradient vector $\nabla\phi$ are parallel in an isotropic medium. Since gradient vector $\nabla\phi$ is always normal to constant-ϕ surface, it is obvious that the specific discharge vector q is also normal to constant-ϕ surface *in an isotropic medium.*

The last assertion is further exemplified by the illustration shown in Figure 2.16, which represents a uniform velocity field through an *isotropic* porous medium contained in a circular tube. Since the velocity vector q is

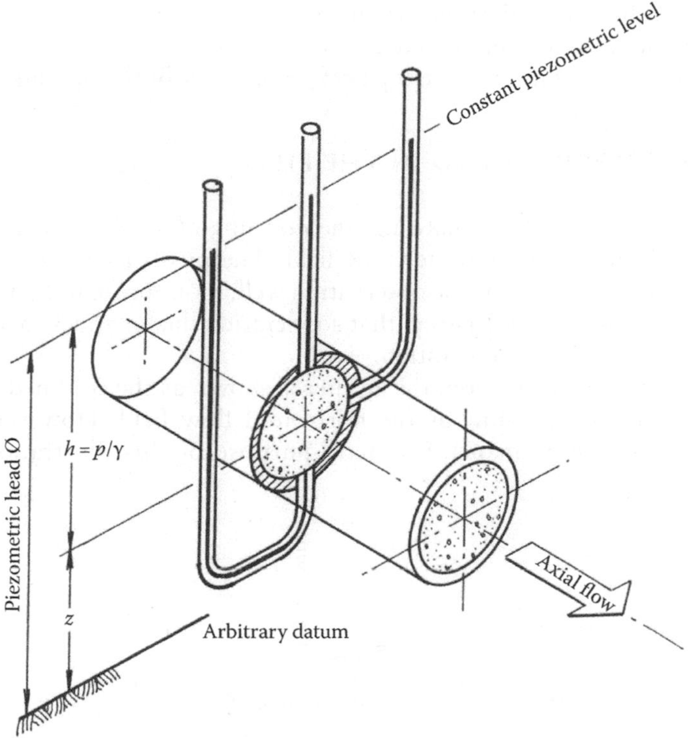

Figure 2.16 An orthogonal section of the axial flow represents a constant Ø-surface.

parallel to the axis of the tube, any orthogonal cross section of the tube represents a constant-ϕ surface, where the sum $\phi = h + z$ remains constant. The middle section with three piezometers installed at the circumference, as shown in Figure 2.16, represents a typical constant-ϕ surface. Since this section represents a surface where ϕ is constant, the three piezometric heads represent one and the same elevation as illustrated in the figure. In fact, at any point in this section (including the center of the circular cross section), the piezometric head is the same as indicated by any of the three piezometers installed at the circumference. This last observation has some further practical ramification. For instance, to measure the piezometric head at any interior point of the flow, one needs to install a piezometer at any convenient point on the circumference of the orthogonal section, without disturbing the flow in the interior.

Finally, as a concluding remark, let us emphasize the fact that the preceding argument does not depend on the inclination of the tube. It is only for convenience of illustration that the axis is shown horizontal in the figure.

Otherwise, the axis of the tube may incline downward, upward, or be horizontal, so long as the piezometric head is measured vertically from the arbitrarily chosen datum to the top surface of water in the piezometer.

2.5 PIEZOMETRIC HEAD IN THE FIELD

Figure 2.17 illustrates schematically the meaning of a piezometric head at any point P in the groundwater flow field. The piezometer shown in the sketch is also referred to as the observation well, monitoring well, or simply as the standpipe. It is understood that acceleration due to gravity acts vertically downward in this illustration.

For the ease of illustration, the aquifer is shown as the confined aquifer and the groundwater flow as the horizontal flow field. However, none of these facts are germane. The main purpose of this illustration is to

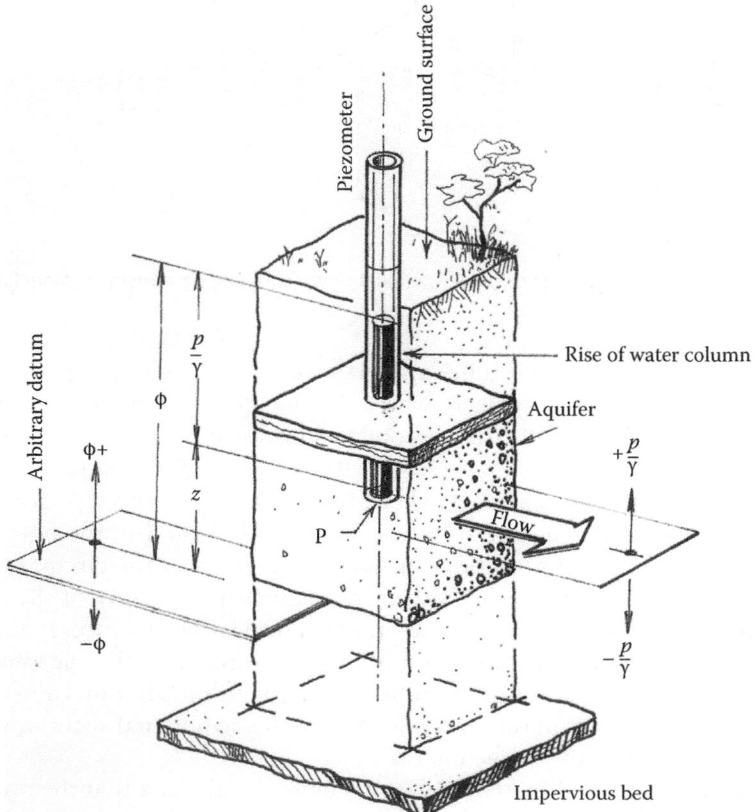

Figure 2.17 Definition sketch for piezometric head in the field.

highlight geometrically the significance of various terms such as piezometric head (ϕ), elevation head (z), and the pressure head (p/γ). All these quantities are scalars, with values positive, negative, or zero. The positive quantities are shown upward. With the exception of pressure head, these quantities are measured with respect to an arbitrary datum, which is generally taken as a geodetic bench mark, the horizontal impervious bedrock, or any convenient level. The pressure head (p/γ) is, however, measured with respect to a horizontal plane passing through the point P. A zero pressure head corresponds with the atmospheric pressure. Thus, a negative value of pressure head implies a suction pressure, which is not relevant in the case of groundwater movement. It is illustrated for completeness, and is meaningful only in the case of vadose water, where pore pressure is subatmospheric.

2.6 ILLUSTRATIVE PROBLEMS

2.1 The following figure illustrates two configurations of piezometers tapped to a pipe carrying water at a pressure of 10.0 kPa (kilo-Pascal) at the center, C, of the cross section. Knowing the pressure at the center is the same as the pressure at point P, find the rise of water column in the two piezometers shown in (a) and (b).

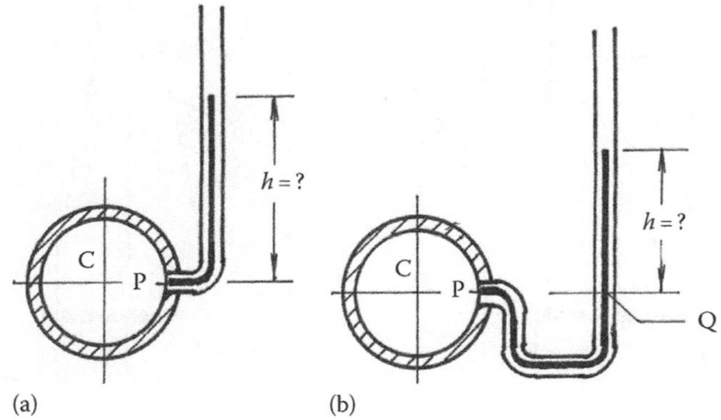

Solution: It is given that $p_C = p_P = 10$ kPa = 10 kN/m^2.

The height of water column in piezometers (a) and (b) with respect to point P is the same, because the pressure at Q is the same as the pressure at P. The height is given by the following formula:

$$h = \frac{p_C}{\gamma} = \frac{10 \text{ kN/m}^2}{9.810 \text{ kN/m}^3} = 1.019 \text{ m}$$

where γ represents the unit weight of water and is found from the following formula:

$$\gamma = (\text{Density of water})(\text{acceleration due to gravity})$$

Thus,

$$\gamma = \left(1000\ \text{kg/m}^3\right)\left(9.81\ \text{m/s}^2\right) = 9.810\ \text{kN/m}^3$$

where
$N \equiv 1.0\ \text{kg} \cdot \text{m/s}^2$.

2.2 Two arrangements of automatic gate opening are illustrated in the following figure. It is required that the circular plate (gate) should open when the water height reaches 2 m above the plate. Find the counterweight in both cases so that the gate opening requirement is met. Ignore the weight of the circular plate in comparison to the hydrostatic pressure force. Assume the string-pulley system is frictionless.

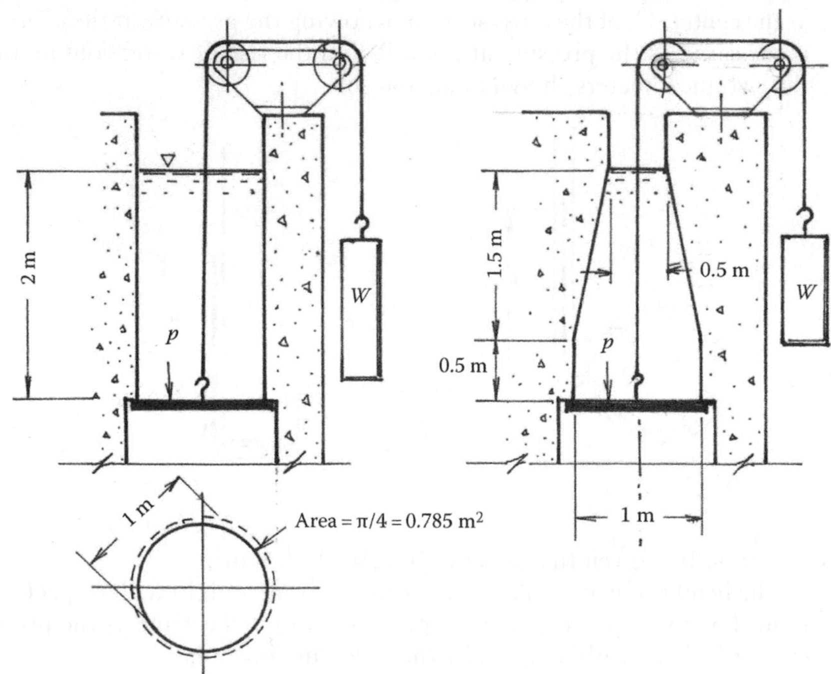

Solution: The pressure p acting on the circular plate is given by the following equation:

$$p = \gamma h = \left(9.810\,\text{kN/m}^3\right)\left(2.0\,\text{m}\right) = 19.62\,\text{kN/m}^2 \tag{IP2.2.1}$$

The force due to pressure on the circular plate is given by the following equation:

$$\begin{aligned}\text{Force, } F &= \text{Pressure} \times \text{Area} \\ &= 19.62\ \text{kN/m}^2 \times 0.785\ \text{m}^2 = 15.40\ \text{kN}\end{aligned} \tag{IP2.2.2}$$

This force should equal the counterweight, that is,

$$W = F = 15.40\ \text{kN} \tag{IP2.2.3}$$

The magnitude of the counterweight is the same in both cases, because force due to pressure on the circular plate is the same. This force depends only on the magnitude of pressure and the magnitude of area on which the pressure is acting. *It (the pressure force) has little to do with the weight of water supported by the circular plate.* In fact, the weight of water supported by the circular plate in part (b) of the figure is only 11.09 kN, which is much less than 15.40 kN—the force due to pressure.

2.3 The dry density of soil sample with porosity 0.4 is found to be 1650 kg/m^3. Find the void ratio and the specific gravity of the sample.

Solution: The void ratio is defined as

$$\text{Void ratio, } e = \frac{\text{Volume of voids}}{\text{Volume of soil skeleton}} = \frac{V_v}{V_s} = \frac{V_v}{V - V_v} \tag{IP.2.3.1}$$

or

$$e = \frac{V_v / V}{1 - V_v / V} = \frac{n}{1 - n} \tag{IP.2.3.2}$$

where n is the porosity. Substituting the value of n, we obtain the following:

$$e = \frac{0.4}{1.0 - 0.4} = 0.6667 \tag{IP.2.3.3}$$

Referring to the following figure, the dry density is given by the following expression:

$$\rho_d = \frac{M_{solid}}{V} = \frac{M_{solid}}{V_s(1+e)} \tag{IP.2.3.4}$$

Segregated phases

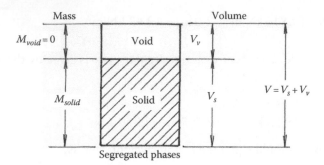

Thus, the mass of solid, M_{solid}, is given by the following expression:

$$M_{solid} = V_s(1+e)\rho_d \tag{IP.2.3.5}$$

Likewise, the mass of water occupying the space V_s is given by the following:

$$M_{water} = V_s\rho_w \tag{IP.2.3.6}$$

Thus, by definition, the specific gravity of the solid grains is given by the ratio of the two masses, as follows:

$$G_s = (1+e)\frac{\rho_d}{\rho_w} \tag{IP.2.3.7}$$

where G_s stands for the specific gravity of the solid grains. Substituting the values in the preceding equation yields

$$G_s = (1+0.6667)\frac{1650\ \text{kg/m}^3}{1000\ \text{kg/m}^3} = 2.75 \tag{IP2.3.8}$$

In the preceding equation, 1000 kg/m^3 represents the density of water in SI system.

2.7 EXERCISES

2.1 Find the pressure in kN/m^2 at a point 10 m below the surface of an ocean. Ignore the compressibility of water and the atmospheric pressure at the surface of the ocean. Density of saltwater may be taken as 1.025 g/cm^3.

2.2 Repeat Exercise 2.1 for a freshwater lake. The density of freshwater may be taken as 1.000 g/cm^3.

2.3 An open tank contains 6.0 m of water overlain with 3.0 m of oil. Find the pressure at the bottom as well as at the interface between the two liquids. Assume the specific gravity of oil as 0.8.

2.4 The following figure illustrates two different sets of situations. In part figure (a), the liquid flowing through the pipe is freshwater and it is the same liquid used in the piezometer to measure the pressure. Likewise, the liquid, both in the pipe and the piezometer in part figure (b), represents saltwater. If the density of saltwater is 1.025 g/cm³ and the pressures in both pipes are same, prove the following equation:

$$\frac{h_f - h_s}{h_s} = \frac{1}{40}.$$

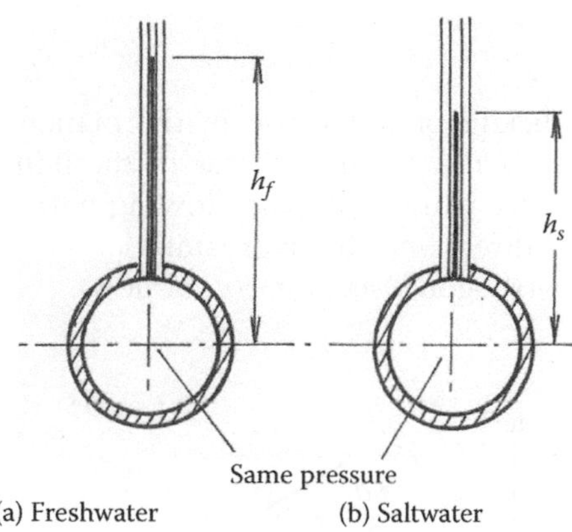

2.5 An open container contains saturated soil, with specific weight 18 kN/m³, to a depth of 2 m (see the following figure). Find the effective pressure p_e at the bottom of the container. Assume the saturated soil behaves like liquids and is in a state of hydrostatic equilibrium.

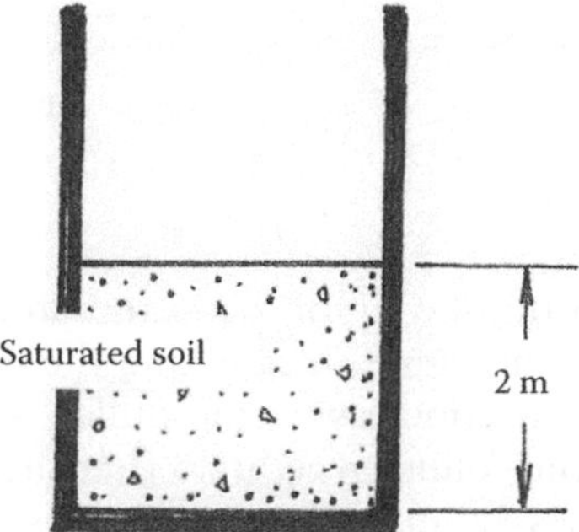

2.6 Repeat Exercise 2.5 if the saturated soil is overlain by water of depth 2 m, as shown in the following figure.

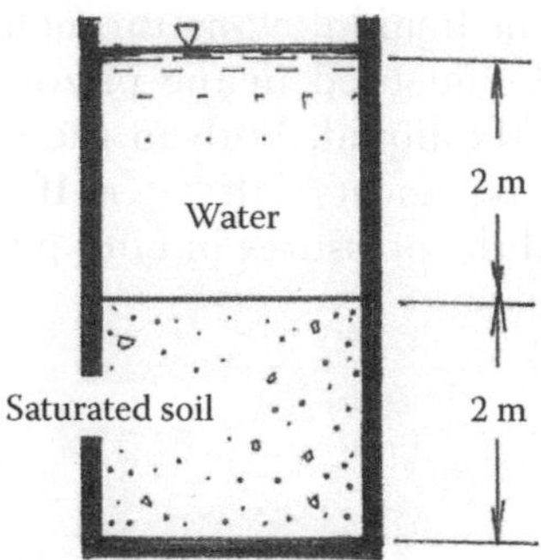

2.7 Determine the height of water rise in the manometer shown in the following figure, if the depth of water in the flume is 1.0 m. (*Hint*: Although the water is not stagnant [flowing with velocity 1.0 m/s in the longitudinal direction], the hydrostatic pressure variation may be assumed along orthogonal section, *nn*, of flow).

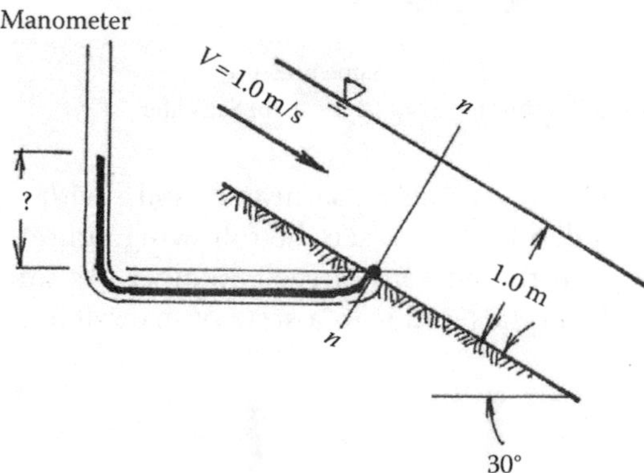

2.8 A longitudinal section of a flume with an automatic plate opening is shown in the following figure. Neglecting the weight of the plate and assuming massless frictionless string-pulley system, determine the magnitude, *W*, of the counterweight so that the plate opens when the depth of flow reaches 3.0 m. (*Hint*: See Exercise 2.7.)

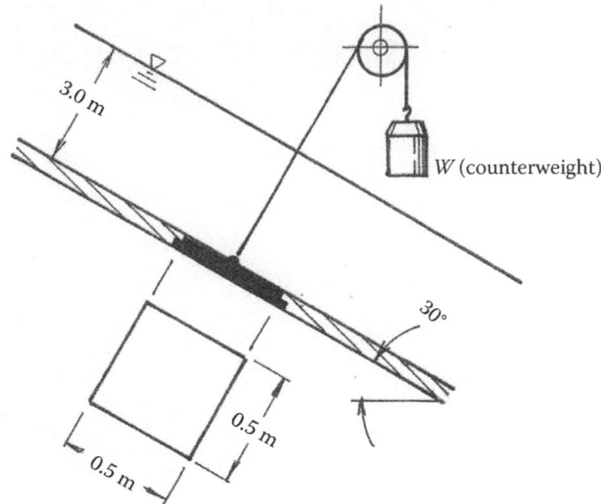

SUGGESTED READINGS

Related to hydrology and the mechanics of groundwater flow in general:

Bear, J. 1979. *Hydraulics of Groundwater*, McGraw-Hill, Inc., New York.

Bear, J. 1988. *Dynamics of Fluids in Porous Media*, Dover Publications, Inc., New York.

Bras, R. L. 1990. *Hydrology: An Introduction to Hydrologic Science*, Addison-Wesley, Reading, MA.

Eagleson, P. S. 1970. *Dynamic Hydrology*, McGraw-Hill, Inc., New York.

Fetter, C. W. 2001. *Applied Hydrogeology*, 4th edn., Prentice Hall, Inc., Upper Saddle River, NJ.

Freeze, R. A. and J. A. Cherry. 1979. *Groundwater*, Prentice Hall, Englewood Cliffs, NJ.

Hubbert, M. K. 1957. *Darcy's Law and the Field Equations of the Flow of Underground Fluids*, Bulletin de l'Association Internationale d' Hydrologie Scientifique, n° 5, 1957. Also, Publication No. 104, Shell Development Company, Exploration and Production Research Division, Houston, TX.

Polubarinova-Kochina, P. Ya. 1962. *Theory of Ground Water Movement*, translated from the Russian by J. M. Roger De Wiest, Princeton University Press, Princeton, NJ.

Todd, D. K. 1959. *Ground Water Hydrology*, John Wiley & Sons, Inc., New York.

Related to mechanics and soil mechanics:

Housner, G. W. and D. E. Hudson. 1959. *Applied Mechanics Dynamics*, 2nd edn., D. Van Nostrand Company, Inc., Princeton, NJ.

Lambe, T. W. and R. V. Whitman. 1969. *Soil Mechanics*, John Wiley & Sons, Inc., New York.

Peck, R. B., Hanson, W. E., and T. H. Thornburn, 1974. *Foundation Engineering*, 2nd edn., John Wiley & Sons, Inc., New York.

Terzaghi, K. 1943. *Theoretical Soil Mechanics*, John Wiley & Sons, Inc., New York.

Chapter 3

Field equations of flow through a porous medium

In this chapter, we intend to lay down the mathematical foundations for the analysis of groundwater movement through a porous medium. For convenience, we shall assume that the fluid and the porous medium are incompressible, and the flow is in a state of steady equilibrium.

3.1 DARCY'S LAW

Henry Darcy, a French hydraulic engineer, conducted a series of experimental studies to investigate the quantitative behavior of flow of water through homogeneous filters of sands. A schematic of his experimental apparatus is shown in Figure 3.1a. Using this apparatus, he investigated the relationship between discharge, cross-sectional area of flow, difference between the manometer heights, and the length of seepage of flow. In his experimental setup, water moved vertically down through the sand column, and mercury was used as the manometer fluid. However, in reporting the experimental observation, the mercury manometer heights were converted into equivalent water heights. An excellent account of Darcy's experimental study is given by Hubbert (1957). The illustrations in Figure 3.1 are adapted from this reference.

It is, however, more instructive to modify Darcy's apparatus to permit the possibility of investigating the flow through an inclined isotropic, homogeneous column of sand as shown in Figure 3.2. This modification is particularly significant in clearly defining each variable and emphasizing the importance of pressure, or lack of it, in determining the direction of flow.

An *isotropic, homogeneous* sample of sand is packed inside the cylindrical tube of cross-sectional area A and is retained by two porous plugs (of negligible resistance), as shown in Figure 3.2. The two ends of the tube are connected to two constant-head reservoirs. The elevation of water in the reservoir can be adjusted to vary the head difference across the sand column. Two piezometric tubes, δx distance apart, are connected to the cylinder to observe the piezometric head difference between Point 1 and Point 2. The water flows through the sand column from the upper reservoir to the

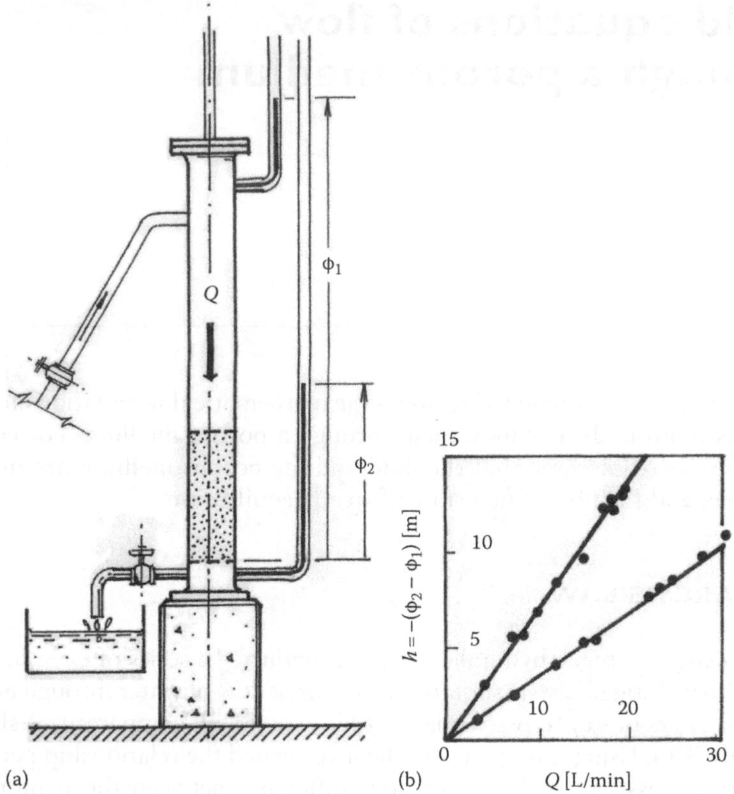

Figure 3.1 (a) Schematic of Darcy's apparatus and (b) proportional relationship between discharge Q and head difference, $\phi_1 - \phi_2$. (Adapted from Hubbert, M.K. Darcy's Law and the field equations of the flow of underground fluids, *Bulletin de l'Association Internationale d'Hydrologie Scientifique*, n° 5, 1957.)

lower for a sufficient duration of time to ensure a saturated, steady-state condition. For such a steady-state condition, and a fixed—but arbitrary—angle of inclination, α, of the tube, the following three empirical observations can be made:

1. If the length δx and the cross-sectional area of flow A are kept constant, the discharge Q varies in proportion to the piezometric head difference, $(\phi_1 - \phi_2)$.
2. If the cross-sectional area A and the head difference $(\phi_1 - \phi_2)$ are kept constant, the discharge Q varies in an inverse proportion to seepage length δx.
3. If the piezometric head difference $(\phi_1 - \phi_2)$ and the seepage length δx are kept constant, the discharge Q varies in proportion to the cross-sectional area A.

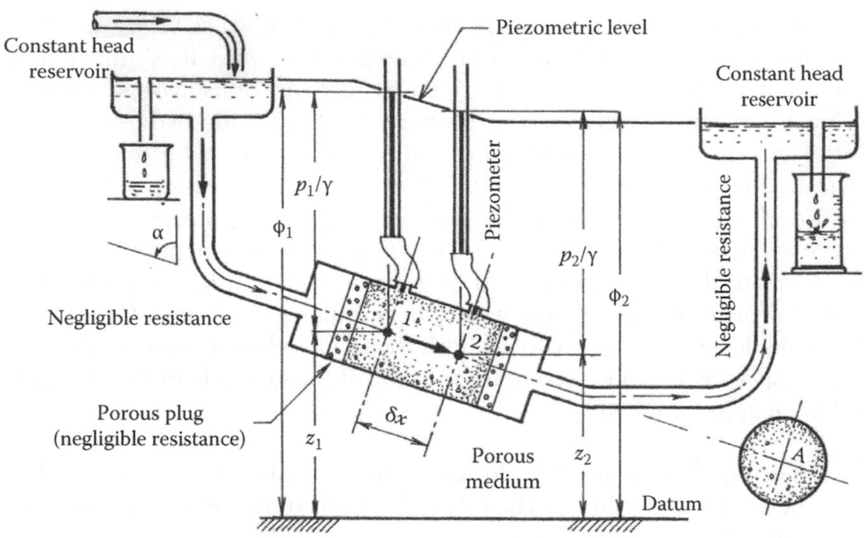

Figure 3.2 Definition sketch for Darcy's law.

The aforementioned proportionality relationships can be replaced by the following equality relationship:

$$Q = KA\frac{(\phi_1 - \phi_2)}{\delta x} = -KA\frac{(\phi_2 - \phi_1)}{\delta x} = -KA\frac{\delta\phi}{\delta x} \tag{3.1}$$

where K is a constant of proportionality. The preceding relationship can also be written as

$$q \equiv \frac{Q}{A} = -K\frac{\delta\phi}{\delta x} \tag{3.2}$$

where q is the specific discharge, which represents the discharge per area and has the dimension of velocity [L/T] (see Section 2.3). It may be emphasized that the cross-sectional area A is at right angle to the main flow direction. If we let $\delta x \to 0$, Equation 3.2 reduces to the following ordinary differential equation:

$$q = -K\frac{d\phi}{dx} = K\frac{-d\phi}{dx} = Ki \tag{3.3}$$

where $i \equiv -d\phi/dx$, which is commonly called the *hydraulic gradient*, and it represents a negative slope of the piezometric level when the axis of the tube is horizontal, that is, $\alpha = 90°$ (Figure 3.2). Since the hydraulic gradient is dimensionless [L/L], it is evident from Equation 3.3 that K has the same dimension as that of velocity [L/T]. Furthermore, in a flow under unit hydraulic gradient, q is numerically equal to K. In other words, a larger value of K means a larger specific discharge q, under a unit hydraulic gradient. Thus, this constant of proportionality can be assigned a physical significance: It quantifies *the ease with which the porous medium permits the fluid to flow*. It is therefore called the *coefficient of permeability*. Among hydrogeologists, the term *hydraulic conductivity* is more popular. It is easy to speculate that for a given fluid at a given temperature, the value of K should depend on the physical nature of porous medium only. Otherwise, K should depend on the nature of both the fluid and the porous medium.

To dispel any preconceived notions, some of the salient features of Figure 3.2 are emphasized here:

1. The piezometric head is plotted vertically, only for those points that lie on the center line of the tube. The locus of this plot represents the piezometric level in Figure 3.2.
2. The fluid in the region from the upper reservoir to the upstream end of the porous sample is assumed to be in a *quasi-static* condition. In this region, the piezometric level is shown constant in the figure. The same is also implied for the fluid in the region downstream from the lower end of the sample. This assumption is not necessary and is used only to simplify the illustration. What is important is the fact that ϕ_1 and ϕ_2 do indeed represent the piezometric levels at Point 1 and Point 2, respectively.
3. Furthermore, the piezometric level does drop *linearly* along the sample, because of the *homogeneity* of the porous medium. This linear drop has little to do with the smallness or largeness of the sample length δx.

Referring to the empirical development leading to Equation 3.3, it is worth emphasizing the fact that the derivation of this equation does not depend on the inclination angle α of the cylindrical tube. In fact, the conclusion is true for any value of α. With $\alpha = 0$, the experimental setup in Figure 3.2 essentially conforms to the apparatus used by M. Henry Darcy, as reported in 1856 (Hubbert, 1957). Equation 3.3 represents a proportional relationship between the specific discharge q and the hydraulic gradient $-d\phi/dx$. Figure 3.1 also indicates the proportionality between discharge and the drop of piezometric head across the homogeneous sand column, as experimentally observed by Darcy. Thus, Equation 3.3 is appropriately known as Darcy's law in his honor. It is analogous to Ohm's law of electric resistance, Fourier's law of heat conduction, or Fick's law of diffusion. Darcy's law has provided the greatest stimulus to the analytical development of flow through the porous media; and, it has indeed placed the subject of groundwater flow, *at par*, with those fields that are governed by the Laplace equation in physics.

Finally, to observe the role of pressure in determining the direction of flow, the following facts may be noted from the figure: (1) when the axis of the tube is parallel to the piezometric level, the pressure head p_1/γ is equal to p_2/γ, that is, the flow through the porous medium takes place despite the absence of pressure gradient; (2) when the axis of the tube is inclined at a steeper inclination than the piezometric level, the flow through the porous medium takes place in an adverse direction of pressure, that is, $p_1/\gamma < p_2/\gamma$; and (3) when the axis of tube is at a flatter slope than the dip of the piezometric level, the flow takes place in the direction of decreasing pressure. Thus, in groundwater flow, the direction of pressure gradient, in itself, has little to do with the direction of specific discharge q. It is the negative gradient of the piezometric head (same as the total energy head, if the velocity head is negligible) that determines the direction of specific discharge q in a porous medium.

3.1.1 Generalization of Darcy's law

We extend Darcy's law to cover three-dimensional flows through porous media. For simplicity of thought, we would assume the porous medium to be *homogenous* and *isotropic*. If the medium is isotropic, the coefficient of permeability, K, does not depend on the direction of flow. Thus, as a generalization of Darcy's law (Equation 3.3) for three-dimensional flows in an *isotropic* porous medium, the following form may be proposed:

$$q_x = -K\frac{\partial \phi}{\partial x} \tag{3.4a}$$

$$q_y = -K\frac{\partial \phi}{\partial y} \tag{3.4b}$$

$$q_z = -K\frac{\partial \phi}{\partial z} \tag{3.4c}$$

where q_x, q_y, q_z represent the rectangular components of the specific discharge vector, q, in an arbitrary, three-dimensional, right-handed Cartesian coordinate system. The preceding three scalar equations can also be represented by a single vector equation:

$$q = -K\nabla\phi \tag{3.5}$$

By tradition, the coefficient of permeability K, like many other coefficients in physics, is represented by a positive scalar quantity. If K is a

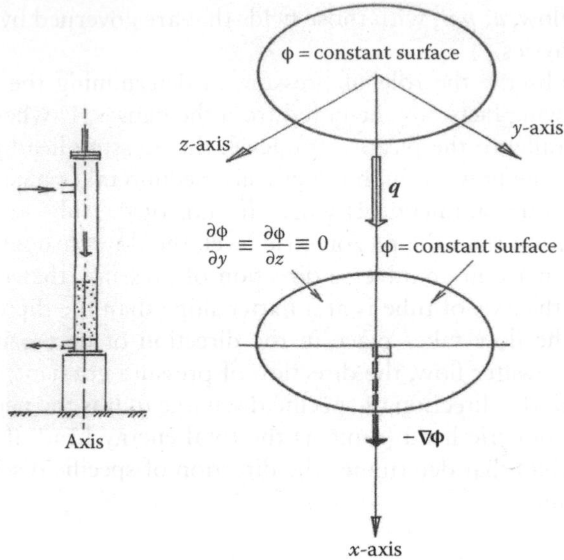

Figure 3.3 Definition sketch for one-dimensional flow parallel to the axis of sand column in Darcy's apparatus.

positive scalar, then the last equation indicates that the two vectors q and $-\nabla\phi$ are parallel and have the same sense of direction in an *isotropic medium*.

Furthermore, if the generalization of Darcy's law, as proposed in Equations 3.4, is true, it must reduce to Equation 3.3 for one-dimensional uniform flow. To be concrete, let the specific discharge vector q point downward; and, since the orientation of the coordinate system is arbitrary, let the x-axis also point downward as shown in Figure 3.3. With this choice of axes, $q_y \equiv q_z \equiv 0$. Since the vectors q and $-\nabla\phi$ are parallel in an *isotropic* medium, it follows that any horizontal surface (normal to q) must represent a constant ϕ-surface. Thus, the partial derivatives of ϕ with respect to y and z must be zero at all points, that is,

$$\frac{\partial\phi}{\partial y} = \frac{\partial\phi}{\partial z} \equiv 0$$

Consequently, the set of three equations (3.4) reduces to a single equation and two identities as follows:

$$q = q_x = -K\frac{d\phi}{dx} = Ki$$

$$q_y \equiv 0(\text{by choice}); \quad -K\frac{\partial\phi}{\partial y} \equiv 0(\text{by necessity})$$

$$q_z \equiv 0(\text{by choice}); \quad -K\frac{\partial\phi}{\partial z} \equiv 0(\text{by necessity})$$

The components q_y and q_z are identically equal to zero because of the choice of coordinate system with respect to the direction of specific discharge vector $\boldsymbol{q}$. However, the partial derivatives of ϕ with respect to y and z are identically equal to zero as a necessary consequence of the isotropy of the medium. Thus, the preceding statements are identical to the original statement of Darcy's law as embodied in Equation 3.3. Hence, for a three-dimensional *isotropic* medium, the generalization of Darcy's law, as given in Equation 3.4, appears acceptable.

Finally, since the generalized Darcy's law involves three equations but four unknowns, q_x, q_y, q_z, ϕ, we need another equation to close the mathematical formulation. This fourth equation is provided by the conservation of mass (or continuity) of fluid flow. This last equation is the subject matter of Section 3.2.

Before we conclude this section, it is worthwhile to look back on Equation 3.5, which can be rearranged in the following manner:

$$\frac{q}{K} = -\nabla\phi \tag{3.6}$$

The left-hand side of the preceding equation represents the dimensionless velocity vector (because scalar K equals velocity under a unit hydraulic gradient) and the right-hand side represents the negative gradient of a scalar function $\phi(x, y, z)$. It is traditional in applied mathematics and mechanics to call such a scalar function as the potential function. Thus, in an *isotropic medium (not necessarily homogeneous)*, the piezometric head does represent a potential function. This potential function $\phi(x, y, z)$ can therefore be called the *dimensionless specific discharge potential*. However, for an *isotropic* medium that is also *homogeneous* in nature, Equation 3.5 can be written as

$$q = -\nabla(K\phi) \tag{3.7}$$

This, incidentally, shows that the velocity vector q can also be obtained from a scalar, potential function $\Phi \equiv (K\phi)$. Thus, for a specific discharge vector $q(x, y, z)$ in a three-dimensional medium, which is *isotropic* as well as *homogeneous*, there exists a potential function, $K\phi(x, y, z)$, which may incidentally be called the *specific discharge potential*. For further discussion on potential functions, please see Section 2.1.3.2.

3.1.2 Laboratory determination of coefficient of permeability

Since there is no reliable information in general on the values of coefficient of permeability, it becomes necessary to determine this coefficient experimentally, either in the laboratory, or in the field. The field determination of aquifer characteristics—as this is commonly called—is discussed in Chapter 7. The laboratory determination is discussed here and is essentially performed using two types of permeameters: the one is called the constant head permeameter, and the other the variable head permeameter. These two types are shown in Figures 3.4 and 3.5.

Constant head permeameter (*Figure 3.4*): This apparatus is used for soil of high permeability, where the discharge through the soil sample can be experimentally observed in a reasonable duration with a reasonable accuracy. This method gives reliable results for clean sands and gravels. For the computation of K, the following equation is used (which is simply a restatement of Darcy's law):

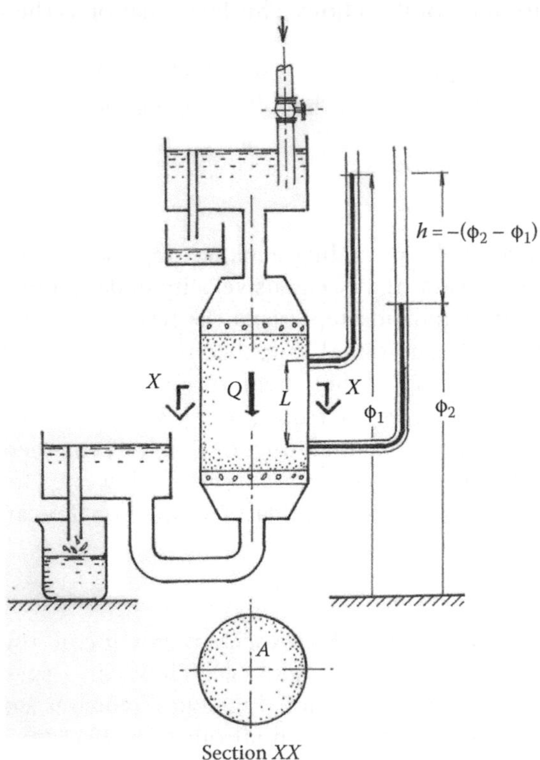

Figure 3.4 A constant head permeameter.

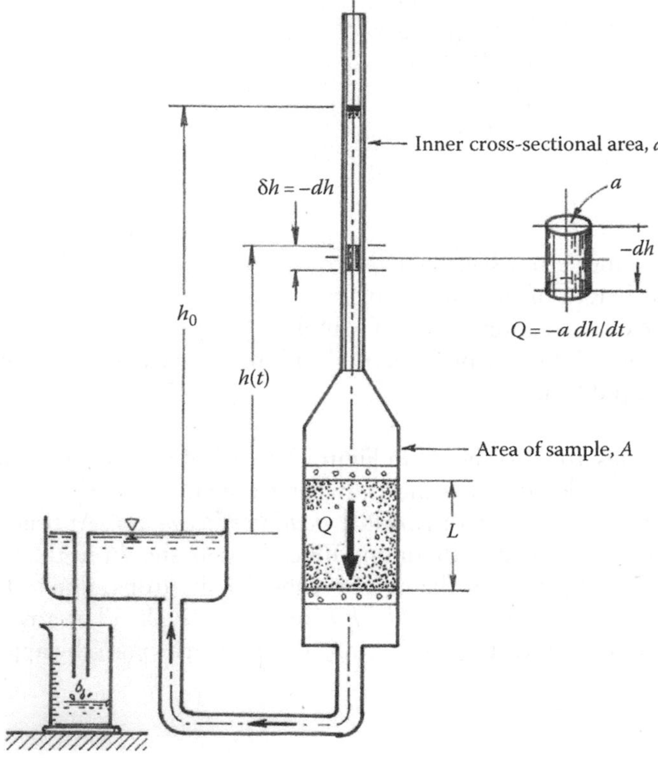

Figure 3.5 A variable head permeameter.

$$K = \frac{QL}{Ah} \tag{3.8}$$

where

- Q is the observed discharge
- L is the length of the soil sample
- A is the cross-sectional area
- $h = -(\phi_2 - \phi_1)$ is the piezometric head difference across the sample

These variables are also shown in Figure 3.4. All of the variables shown on the right-hand side are experimentally determined.

Variable-head permeameter (*Figure 3.5*): This apparatus is used for soils of low permeability. The experimental setup is shown in the sketch, which allows considerable variations of head, and time of flow, to account for the variability of K values. To determine K, the following equations can be used:

$$K = \frac{-aL}{At}\ln\left(\frac{h}{h_0}\right) = \frac{aL}{At}\ln\left(\frac{h_0}{h}\right) \tag{3.9a}$$

$$K = 2.3\frac{-aL}{At}\log_{10}\left(\frac{h}{h_0}\right) = 2.3\frac{aL}{At}\log_{10}\left(\frac{h_0}{h}\right) \tag{3.9b}$$

where

- a is the inner cross-sectional area of the narrow tube
- L is the length of the soil sample
- A is the cross-sectional area of sample
- $h(t) = -(\phi_2 - \phi_1)$ is the piezometric head difference across sample at any observed time t

These variables are also shown in Figure 3.5. All of the variables shown on the right-hand side are experimentally determined.

Derivation of equation for variable-head permeameter: At time $t = 0$, the permeameter is charged with an initial head h_0, as shown in the figure. As time proceeds, the head in the narrow tube slowly drops, due to flow into the constant-head reservoir. Let $h = h(t)$ represent the head at any time t as shown in the sketch. At this instant, the discharge through the apparatus is

$$Q(t) = -a\frac{dh}{dt} \tag{3.10}$$

The negative sign means that the differential of column height, dh, decreases as the differential of time dt increases. In other words, Q is considered positive when the head $h(t)$ in the narrow tube falls. From the geometry of the differential element shown in the sketch, it is clear that the expression on the right-hand side of Equation 3.10 represents the rate of change of volume with respect to time. This, of course, is the definition of discharge. From Darcy's law, the discharge through the apparatus at any time t is given by

$$Q(t) = KA\frac{h}{L} \tag{3.11}$$

where h indeed represents the head difference, $-(\phi_2 - \phi_1)$, across the porous sample of length L. (Why? Prove it!) Thus, combining the last two equations yields the following ordinary differential equation:

$$\frac{dh}{dt} + K\frac{hA}{La} = 0 \tag{3.12}$$

The preceding equation is separable and on integration yields the following expression for $h(t)$:

$$\ln h(t) = -\left[K\left(\frac{A}{aL}\right)\right]t + C \tag{3.13}$$

To eliminate the arbitrary constant, C, of integration, let $h(t)=h_0$ at time $t=0$ in the preceding equation. Thus,

$$\ln\left(\frac{h_0}{h}\right) = \left[K\left(\frac{A}{aL}\right)\right]t$$

which finally yields the two expressions for K, given in Equations 3.9a and 3.9b.

3.2 CONSERVATION OF MASS, OR CONTINUITY, EQUATION

As mentioned in the previous section, we need another equation to complete the mathematical formulation for the determination of four unknowns, q_x, q_y, q_z, ϕ. The required equation is provided by considering the conservation of mass inside a small (infinitesimal) parallelepiped (Figure 3.6). For this purpose, we assume that the specific discharge $q(x, y, z, t)$ and fluid density $\rho(x, y, z, t)$ are in general functions of space and time.

Now, let us consider two faces normal to the x-axis indicated by Face 1 and Face 2, having a nominal cross-sectional dimension $\delta y \cdot \delta z$, as shown in Figure 3.6. Let q and ρ represent the specific discharge vector and the density of fluid, respectively, at the center of Face 1. Let us first consider the transport of mass across Face 1 *into* the parallelepiped during a small (infinitesimal) duration of time δt, as follows:

$$\rho q_x (\delta y \delta z)\delta t \tag{3.14a}$$

It may be emphasized that only the normal component q_x of q is instrumental in transporting the mass across Face 1. Furthermore, the quantity $q_x(\delta y \delta z)\delta t$ indeed represents the volume of fluid flow across Face 1 during time interval δt, *despite the fact that only part of the nominal area* $(\delta y \delta z)$ *is available for flow due to the presence of granular material.* The last assertion is implicit in the definition of the specific discharge vector q. Likewise, the transport of mass across Face 2 *out of* the parallelepiped is

$$\left[\rho q_x + \delta(\rho q_x)\right](\delta y \delta z)\delta t \tag{3.14b}$$

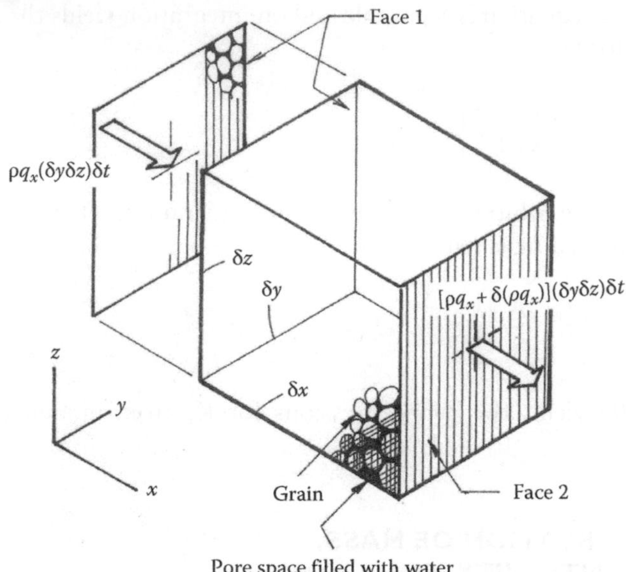

Figure 3.6 Definition sketch for conservation-of-mass equation.

Thus, the net gain of mass inside the parallelepiped during the time interval δt, due to the mass flow across the two faces, Face 1 and Face 2, becomes

$$\rho q_x(\delta y \delta z)\delta t - \left[\rho q_x + \delta(\rho q_x)\right](\delta y \delta z)\delta t = -\delta(\rho q_x)(\delta y \delta z)\delta t \qquad (3.14c)$$

In the preceding equation, the expression on the right-hand side can be replaced by a truncated Taylor's series, ignoring higher order terms. Thus, the net gain due to mass flow across Face 1 and Face 2 becomes

$$-\delta(\rho q_x)(\delta y \delta z)\delta t = -\frac{\partial(\rho q_x)}{\partial x}\delta x(\delta y \delta z)\delta t \qquad (3.14d)$$

Similarly, when the mass transport takes place across all (six) faces, the net gain of mass inside the parallelepiped during time interval δt is given by the following:

$$\text{Net gain} = -\left[\frac{\partial(\rho q_x)}{\partial x} + \frac{\partial(\rho q_y)}{\partial y} + \frac{\partial(\rho q_z)}{\partial z}\right]\delta x \delta y \delta z \delta t \qquad (3.14e)$$

Let the mass of fluid inside the parallelepiped at any time be $m(t)$. Since only the pore space can contain fluid, it is evident that for a completely saturated medium this mass should be given by the following:

$$m(t) = \rho(n\delta x\delta y\delta z) \tag{3.14f}$$

where

n is the porosity

$(n\delta x\delta y\delta z)$ is the volume of the total pore space contained by the parallelepiped

The increment of mass $m(t)$ during time interval δt thus becomes

$$\delta m = \frac{\partial(\rho n)}{\partial t}\delta t(\delta x\delta y\delta z) \tag{3.14g}$$

because the volume of the parallelepiped does not change with time. Hence, equating the net gain (Equation 3.14e) to the increment of m (Equation 3.14g) yields the final equation:

$$-\left[\frac{\partial(\rho q_x)}{\partial x} + \frac{\partial(\rho q_y)}{\partial y} + \frac{\partial(\rho q_z)}{\partial z}\right]\delta x\delta y\delta z\delta t = \frac{\partial(\rho n)}{\partial t}\delta t(\delta x\delta y\delta z) \tag{3.14h}$$

Since δx, δy, δz, δt are arbitrary, the preceding equation reduces to

$$-\left[\frac{\partial(\rho q_x)}{\partial x} + \frac{\partial(\rho q_y)}{\partial y} + \frac{\partial(\rho q_z)}{\partial z}\right] = \frac{\partial(\rho n)}{\partial t} \tag{3.15}$$

If the groundwater flow is in a *steady-state* condition—that is, ρ and n do not depend on time—then the preceding equation reduces to the following:

$$-\left[\frac{\partial(\rho q_x)}{\partial x} + \frac{\partial(\rho q_y)}{\partial y} + \frac{\partial(\rho q_z)}{\partial z}\right] = 0 \tag{3.16}$$

Finally, if the fluid is *homogeneous*, we have the so-called continuity equation:

$$-\left[\frac{\partial q_x}{\partial x} + \frac{\partial q_y}{\partial y} + \frac{\partial q_z}{\partial z}\right] \equiv \nabla \cdot q = 0 \tag{3.17}$$

In conclusion, the flow of an incompressible homogeneous fluid in a saturated nonconsolidating aquifer satisfies the continuity equation. These requirements can be more succinctly stated as follows: The groundwater flow in a steady state (which undergoes no change with time) satisfies the continuity equation. This assertion is true irrespective of the nature of porous medium with regard to its isotropy or its homogeneity.

3.3 LAPLACE EQUATION

If we consider the flow of an incompressible fluid in an isotropic, homogeneous nonconsolidating porous medium, the fluid must satisfy the three equations (3.4) of generalized Darcy's law, as well as the continuity equation (3.17). Taken together, these four equations provide the necessary equations for the solution of the four unknowns, q_x, q_y, q_z, ϕ. It is however possible to eliminate the components of specific discharge vector q. To accomplish this, let us substitute the components of the specific discharge vector from the three equations (3.4) into the continuity equation (3.17) as follows:

$$-\left[\frac{\partial}{\partial x}\frac{\partial(-K\phi)}{\partial x}+\frac{\partial}{\partial y}\frac{\partial(-K\phi)}{\partial y}+\frac{\partial}{\partial z}\frac{\partial(-K\phi)}{\partial z}\right]=0 \tag{3.18}$$

Since the medium is homogeneous, the coefficient of permeability K does not depend on the location of a point—it is a constant for the entire medium. Thus, the preceding equation can be written as

$$K\left[\frac{\partial^2\phi}{\partial x^2}+\frac{\partial^2\phi}{\partial y^2}+\frac{\partial^2\phi}{\partial z^2}\right]=0 \tag{3.19}$$

If we further assume that K is not zero, the preceding equation reduces to

$$\left[\frac{\partial^2\phi}{\partial x^2}+\frac{\partial^2\phi}{\partial y^2}+\frac{\partial^2\phi}{\partial z^2}\right]\equiv\nabla^2\phi=0 \tag{3.20}$$

Equation 3.20 is known as the *Laplace equation* and the operator ∇^2 is called the *Laplacian operator.*

Finally, the piezometric head ϕ, at any point in a steady-state flow field of an incompressible, homogeneous fluid in an isotropic, homogeneous, porous medium satisfies the Laplace equation. Thus, the solution of a groundwater flow leads to the solution of the so-called boundary-value problem associated with the Laplace equation, subject to appropriate boundary conditions.

3.4 TWO-DIMENSIONAL ANISOTROPIC MEDIUM AND PERMEABILITY MATRIX (OR TENSOR)

Most geological formations bearing groundwater are to some extent anisotropic and heterogeneous in character. Our objective here is, therefore, to extend Darcy's law to a two-dimensional anisotropic porous medium.

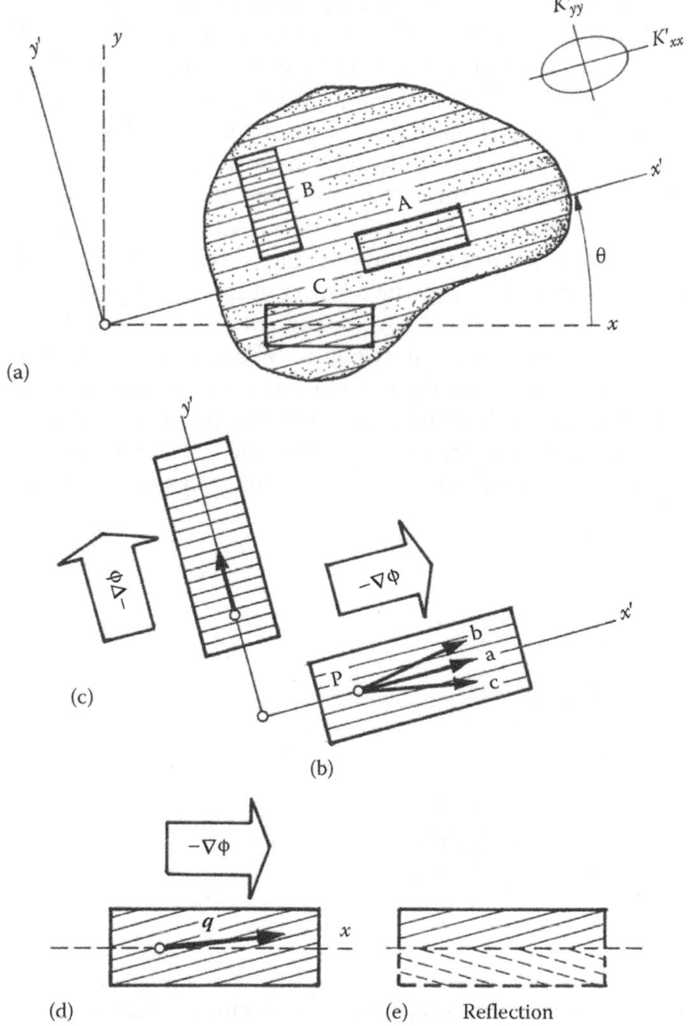

Figure 3.7 (a) A two-dimensional anisotropic medium, (b) hydraulic gradient points in x'-direction, (c) hydraulic gradient points in y'-direction, (d) hydraulic gradient points in x-direction, and (e) asymmetrical specimen.

Let us consider an anisotropic porous medium, as shown in Figure 3.7a. For clarity of perception, such a medium can be visualized to consist of an alternating sequence of two vanishingly thin layers of different permeability. In such a medium, we can recognize two mutually orthogonal lines of symmetry, one along the bedding plane and the other across the bedding plane, as represented by x'-axis and y'-axis in Figure 3.7a. Since the medium is anisotropic, the coefficient of permeability K'_{xx} in x'-direction is, in general, different from the coefficient of permeability K'_{yy} in y'-direction. Let a specimen of this porous medium be cut symmetrically about x'-axis (Specimen A in Figure 3.7a) and subjected to a hydraulic gradient along the x'-axis (Figure 3.7b). This hydraulic gradient will induce a specific discharge vector at point P. From a purely hypothetical viewpoint, the specific discharge vector, $\boldsymbol{q}$, could either point in the x'-direction, or upward, or downward, as shown, respectively, by arrows a, b, c in Figure 3.7b. Since the problem is symmetrical about the x'-axis, the specific discharge vector, $\boldsymbol{q}$, cannot deviate from the x'-direction. A similar argument also holds for Specimen B cut along y'-axis and subjected to a hydraulic gradient in the y'-direction, as shown in Figure 3.7c. Since the situation shown in Figure 3.7b and c is essentially the same as stipulated in Darcy's original experiment—that is, the hydraulic gradient and the specific discharge vector point in the same direction—the following generalization of Darcy's law for an anisotropic medium in x', y'-coordinate system can be proposed:

$$q'_x = -K'_{xx}\frac{\partial\phi}{\partial x'} \tag{3.21a}$$

$$q'_y = -K'_{yy}\frac{\partial\phi}{\partial y'} \tag{3.21b}$$

or in matrix form

$$\begin{Bmatrix} q'_x \\ q'_y \end{Bmatrix} = -\begin{bmatrix} K'_{xx} & 0 \\ 0 & K'_{yy} \end{bmatrix}\begin{Bmatrix} \dfrac{\partial\phi}{\partial x'} \\ \dfrac{\partial\phi}{\partial y'} \end{Bmatrix} \tag{3.21c}$$

where

q'_x and q'_y represent the components of specific discharge vector $\boldsymbol{q}$ along x'- and y'-axis, respectively

$-\dfrac{\partial\phi}{\partial x'}$ and $-\dfrac{\partial\phi}{\partial y'}$ represent the components of the hydraulic gradient $-\nabla\phi$ along x'- and y'-axis, respectively

K'_{xx} and K'_{yy} represent the coefficients of permeability along x'- and y'-axis, respectively

Let us now consider a specimen cut along the x-axis and subject it to a hydraulic gradient acting along the x-axis, as shown in Figure 3.7d. In this case, the porous medium is not symmetrical about the x-axis, because the lower half of the specimen below the x-axis is not the same as the reflection of the upper half into the x-axis (Figure 3.7e). Hence, in this case, we have no a priori reason to believe that the specific discharge vector acts along the x-axis. Consequently, we assume without loss of generality that the specific discharge vector $\boldsymbol{q}$ acts in a direction other than the x-axis (Figure 3.7d). In this case, the hydraulic gradient, despite the fact that it acts in x-direction, can induce a nonzero velocity component, q_y, acting at a right-angle to the gradient vector. Thus, in the x-, y-coordinate system, Darcy's law should have the general linear form:

$$q_x = -\left[K_{xx}\frac{\partial\phi}{\partial x} + K_{xy}\frac{\partial\phi}{\partial y}\right] \tag{3.22a}$$

$$q_y = -\left[K_{yx}\frac{\partial\phi}{\partial x} + K_{yy}\frac{\partial\phi}{\partial y}\right] \tag{3.22b}$$

or in matrix notation

$$\begin{Bmatrix} q_x \\ q_y \end{Bmatrix} = -\begin{bmatrix} K_{xx} & K_{xy} \\ K_{yx} & K_{yy} \end{bmatrix}\begin{Bmatrix} \dfrac{\partial\phi}{\partial x} \\ \dfrac{\partial\phi}{\partial y} \end{Bmatrix} \tag{3.22c}$$

where

q_x and q_y represent components of the specific discharge vector in the x- and y-coordinate system, respectively

$-\dfrac{\partial\phi}{\partial x}$ and $-\dfrac{\partial\phi}{\partial y}$ represent the components of hydraulic gradient in the x- and y-coordinate system, respectively

The elements K_{xx}, K_{xy}, K_{yx}, and K_{yy} of the coefficient matrix physically represent the coefficients of permeability as before. However, among these coefficients, K_{xy} and K_{yx} are rather peculiar, for they represent the cross coupling between the components of specific discharge and hydraulic gradient vectors. For instance, K_{xy} represents *the contribution to x-component of the specific discharge vector by the y-component of the hydraulic gradient vector.* A similar interpretation for K_{yx} is possible by an examination of Equations 3.22. These two coefficients are casually referred to as the *cross permeability coefficients*, and the other two, K_{xx} and K_{yy}, as the *normal permeability coefficients*. Thus, in a reference frame chosen arbitrarily—such as the x, y-coordinate system—without regard to the intrinsic symmetry of the porous medium, Darcy's law requires four coefficients of permeability for relating hydraulic gradient $-\nabla\phi$ to the specific discharge vector $\boldsymbol{q}$. However, as we shall see later, only three of these coefficients are independent, because it turns out that the cross permeability coefficients are equal, that is, $K_{xy} = K_{yx}$.

If the orientation of porous medium is such that the x'-axis coincides with the x-axis ($\theta = 0$ in Figure 3.7a), the coefficient matrix

$$[K] \equiv \begin{bmatrix} K_{xx} & K_{xy} \\ K_{yx} & K_{yy} \end{bmatrix} \tag{3.23}$$

in Equation 3.22c should reduce to

$$[K'] \equiv \begin{bmatrix} K'_{xx} & 0 \\ 0 & K'_{yy} \end{bmatrix} \tag{3.24}$$

given in Equation 3.21c. In other words,

$$K_{xx} \to K'_{xx}; \quad K_{yy} \to K'_{yy} \quad \text{and} \quad K_{xy} = K_{yx} \to 0, \quad \text{as } \theta \to 0$$

This suggests that the elements of $[K]$ matrix, K_{xx}, K_{xy}, K_{yx}, and K_{yy}, should somehow be related to the orientation angle θ. The mathematical entities whose components depend on the rotation (or orientation) of the reference frame constitute the subject matter of tensor analysis. Here, we are not interested per se in this field of mathematics. However, please note that the components of the $[K]$ matrix do constitute a *tensor of second order, or rank.* Since the components of this tensor physically represent the coefficients of permeability, we shall henceforth refer to this as the *permeability tensor,* and the corresponding matrix, the *permeability matrix.* In the particular case when the off-diagonal terms become zero, as in Equation 3.24, the diagonal terms (e.g., K'_{xx} and K'_{yy}) are called the *principal values*, and the corresponding directions, x'-axis and y'-axis, the *principal directions* of the permeability tensor. In our particular case, these principal directions indeed represent the axes of symmetry of the two-dimensional anisotropic porous medium. In the subsequent sections, we shall develop the definite expressions that represent the influence of angle θ on the components of the permeability tensor.

If we choose x', y'-coordinate system as our reference frame, Darcy's law reduces to the mathematical form given in Equations 3.21, where K'_{xx} and K'_{yy} are physical properties of the porous medium. These either are known a priori or can be determined by an experiment. Thus, from a theoretical viewpoint, Equations 3.21 can always serve as a valid starting point for the analysis of flow in an anisotropic medium. However, from a practical viewpoint, such a judicious selection of reference frame may be undesirable—or even impossible in some cases, as in the case of nonuniformly anisotropic medium with changing principal directions of permeability. Thus, we encounter a situation where Darcy's law is known in a particular reference frame, but for practical purposes, we deduce it with respect to an arbitrary reference frame. At the heart of this problem is the fundamental question:

How do the components of a vector, such as the specific discharge vector, or the partial derivatives of a scalar function, transform under rotation of reference system? When the x, y-reference frame is rotated through an angle θ to obtain x', y'-reference frame, the components of discharge vector transform according to the following equation (details are given in Appendix B):

$$\begin{Bmatrix} q'_x \\ q'_y \end{Bmatrix} = \begin{bmatrix} \cos\theta & \sin\theta \\ -\sin\theta & \cos\theta \end{bmatrix} \begin{Bmatrix} q_x \\ q_y \end{Bmatrix} \tag{3.25}$$

The components of the gradient vector in the two reference frames are also related by a similar matrix equation:

$$\begin{Bmatrix} \dfrac{\partial\phi}{\partial x'} \\ \dfrac{\partial\phi}{\partial y'} \end{Bmatrix} = \begin{bmatrix} \cos\theta & \sin\theta \\ -\sin\theta & \cos\theta \end{bmatrix} \begin{Bmatrix} \dfrac{\partial\phi}{\partial x} \\ \dfrac{\partial\phi}{\partial y} \end{Bmatrix} \tag{3.26}$$

where

$$[R] \equiv \begin{bmatrix} \cos\theta & \sin\theta \\ -\sin\theta & \cos\theta \end{bmatrix} \tag{3.27}$$

is the so-called rotation matrix. We should, however, emphasize here that a positive value of θ implies a counterclockwise rotation of x, y-reference system to x', y'-reference system. Substituting the expressions for specific discharge vector and the gradient vector from Equations 3.25 and 3.26 in Equation 3.21c yields

$$[R]\begin{Bmatrix} q_x \\ q_y \end{Bmatrix} = -\begin{bmatrix} K'_{xx} & 0 \\ 0 & K'_{yy} \end{bmatrix} [R] \begin{Bmatrix} \dfrac{\partial\phi}{\partial x} \\ \dfrac{\partial\phi}{\partial y} \end{Bmatrix} \tag{3.28}$$

When the preceding equation is (left) multiplied by the inverse of rotation matrix, it yields

$$\begin{Bmatrix} q_x \\ q_y \end{Bmatrix} = -[R]^{-1} \begin{bmatrix} K'_{xx} & 0 \\ 0 & K'_{yy} \end{bmatrix} [R] \begin{Bmatrix} \dfrac{\partial\phi}{\partial x} \\ \dfrac{\partial\phi}{\partial y} \end{Bmatrix} \tag{3.29}$$

Since the rotation matrix is an orthogonal matrix, its inverse is simply equal to its transpose, that is,

$$[R]^{-1} = \begin{bmatrix} \cos\theta & -\sin\theta \\ \sin\theta & \cos\theta \end{bmatrix} \tag{3.30}$$

Finally, after multiplying out the three square matrices in Equation 3.29, we get

$$\begin{Bmatrix} q_x \\ q_y \end{Bmatrix} = -\begin{bmatrix} \left(K'_{xx}\cos^2\theta + K'_{yy}\sin^2\theta\right) & \left(K'_{xx} - K'_{yy}\right)\sin\theta\,\cos\theta \\ \left(K'_{xx} - K'_{yy}\right)\sin\theta\cos\theta & \left(K'_{xx}\sin^2\theta + K'_{yy}\cos^2\theta\right) \end{bmatrix} \begin{Bmatrix} \dfrac{\partial\phi}{\partial x} \\ \dfrac{\partial\phi}{\partial y} \end{Bmatrix} \tag{3.31}$$

This equation relates the specific discharge vector q to the hydraulic gradient vector $-\nabla\phi$ in a general x, y-reference frame. Thus, Equation 3.31 is the mathematical statement of Darcy's law with respect to x, y-coordinate system, for a two-dimensional anisotropic medium.

Now, comparing Equation 3.31 with Equation 3.22c, we observe that

$$K_{xx} = K'_{xx}\cos^2\theta + K'_{yy}\sin^2\theta \tag{3.32a}$$

$$K_{yy} = K'_{xx}\sin^2\theta + K'_{yy}\cos^2\theta \tag{3.32b}$$

$$K_{xy} = K_{yx} = \left(K'_{xx} - K'_{yy}\right)\sin\theta\,\cos\theta \tag{3.32c}$$

It is evident that the permeability matrix is symmetric, that is, $K_{xy} = K_{yx}$. Moreover, for a given anisotropic porous medium, the elements of the permeability matrix $[K]$ are functions of one independent variable, θ, only, because the principal values, K'_{xx} and K'_{yy}, are the given characteristics of the anisotropic medium. Also, since the elements, K_{xx}, K_{xy}, K_{yy}, are functions of a single variable θ, these elements cannot be chosen arbitrarily. They must always satisfy the restrictions imposed by the preceding three equations. In the following section, we shall describe graphically the restrictions imposed by Equations 3.32 on the elements of the permeability matrix $[K]$.

3.4.1 Mohr's circle

Let us recall the following trigonometric identities:

$$\sin 2\theta = 2\sin\theta\,\cos\theta$$

$$\cos^2\theta = \frac{1+\cos 2\theta}{2}$$

$$\sin^2\theta = \frac{1-\cos 2\theta}{2}$$

Using these identities, we can express the preceding three equations, Equation 3.32, in the following forms:

$$K_{xx} = \frac{K'_{xx}+K'_{yy}}{2} + \frac{K'_{xx}-K'_{yy}}{2}\cos 2\theta \tag{3.33a}$$

$$K_{yy} = \frac{K'_{xx}+K'_{yy}}{2} - \frac{K'_{xx}-K'_{yy}}{2}\cos 2\theta \tag{3.33b}$$

$$K_{xy} = K_{yx} = \frac{K'_{xx}-K'_{yy}}{2}\sin 2\theta \tag{3.33c}$$

For a given value of θ, we can obtain simultaneously the components K_{xx} and K_{xy} from Equations 3.33a and 3.33c. These components, K_{xx} and K_{xy}, can be used to denote the coordinates of a point $P(K_{xx},K_{xy})$ in a plane. We shall refer to this plane as the *permeability plane*, and the coordinates as the *permeability coordinates* of a generic point P. The permeability plane is similar to the usual *x, y*-plane of the analytical coordinate geometry. The abscissa on this plane represents the normal coefficients (K_{xx}, or K_{yy}) and the ordinate, the cross permeability coefficient (K_{xy}). There is, however, a substantive difference between the two planes: The positive direction of the ordinate axis points downward instead of upward (Figure 3.8).The reason for this deviation from the established convention of the analytical coordinate geometry will be explained later. (The treatment given here is very similar to Mohr's graphical analysis of stress tensor for two-dimensional stress field in elasticity).

Taken together, Equations 3.33a and 3.33c can now be viewed as the parametric form of a general plane curve executed in the permeability plane by point P, as the parameter θ is arbitrarily varied. In order to determine an explicit mathematical expression for the shape of this curve, we need to eliminate θ from Equations 3.33a and 3.33c. To accomplish this, we express Equations 3.33a and 3.33c in the following form:

$$\left[K_{xx} - \frac{K'_{xx}+K'_{yy}}{2}\right]^2 = \left[\frac{K'_{xx}-K'_{yy}}{2}\cos 2\theta\right]^2 \tag{3.34a}$$

$$\left[K_{xy}\right]^2 = \left[\frac{K'_{xx}-K'_{yy}}{2}\sin 2\theta\right]^2 \tag{3.34b}$$

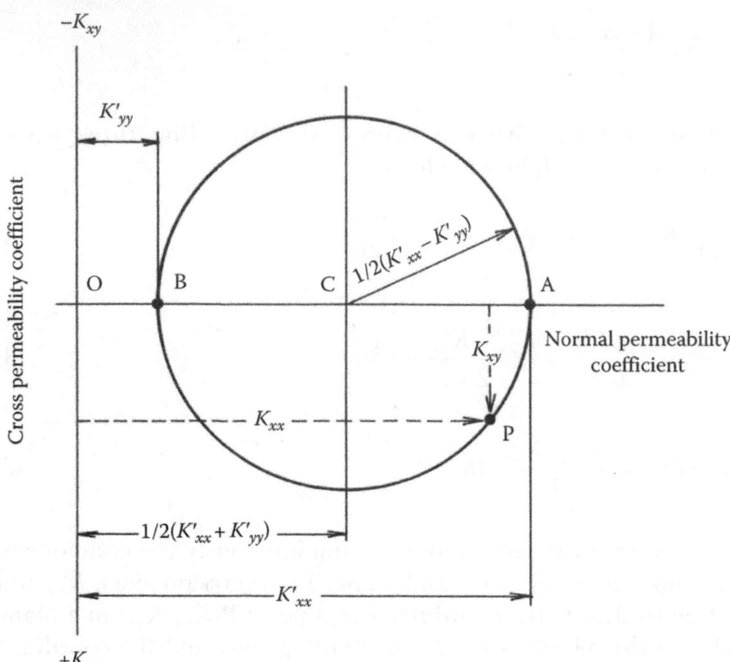

Figure 3.8 Permeability plane and Mohr's circle.

Adding the preceding two equations yields

$$\left[K_{xx}-\frac{K'_{xx}+K'_{yy}}{2}\right]^2+\left[K_{xy}\right]^2=\left[\frac{K'_{xx}-K'_{yy}}{2}\right]^2 \tag{3.35}$$

which is similar in form to the equation of a circle

$$\left[x-x_c\right]^2+\left[y-y_c\right]^2=r^2 \tag{3.36}$$

whose radius is r and the center is at (x_c, y_c). Thus, the locus of point P represents a circle with radius

$$r=\frac{K'_{xx}-K'_{yy}}{2} \tag{3.37}$$

and center on the horizontal axis at a distance

$$x_c=\frac{K'_{xx}+K'_{yy}}{2} \tag{3.38}$$

from the origin, as shown in Figure 3.8. This circle is known as Mohr's circle, named after the prominent German professor of engineering mechanics, who suggested its use in the analysis of stress tensor.

We are now at a point where we can discuss the geometric features of Mohr's circle. For instance, if the principal values K'_{xx} and K'_{yy} of permeability of the anisotropic aquifer are known, we can readily find the radius and the center from Equations 3.37 and 3.38, respectively, and construct the Mohr circle for the anisotropic medium, as shown in Figure 3.8. From this diagram, we observe the following geometric relationships:

$$\overline{OA} = \overline{OC} + r = \frac{K'_{xx} + K'_{yy}}{2} + \frac{K'_{xx} - K'_{yy}}{2} = K'_{xx} \tag{3.39}$$

$$\overline{OB} = \overline{OC} - r = \frac{K'_{xx} + K'_{yy}}{2} - \frac{K'_{xx} - K'_{yy}}{2} = K'_{yy} \tag{3.40}$$

Thus, the abscissas of points A and B represent the principal values of permeability tensor. Using the geometric elements of Mohr's circle, we can express Equation 3.33 in the following form:

$$K_{xx} = \overline{OC} + r\cos 2\theta \tag{3.41a}$$

$$K_{yy} = \overline{OC} - r\cos 2\theta \tag{3.41b}$$

$$K_{xy} = r\sin 2\theta \tag{3.41c}$$

With the help of preceding relationships, a pair of points $P(K_{xx}, K_{xy})$ and $Q(K_{yy}, -K_{xy})$ can be located on the circle, as shown in Figure 3.9a. For lack of a better terminology, we shall refer to this pair of points as the *conjugate pair*.

Before we conclude this section, it is worth emphasizing a few facts:

a. Every point on the Mohr circle—for example, $P(K_{xx}, K_{xy})$ or $Q(K_{yy}, -Kxy)$—can be associated with an independent parameter θ. Since the parameter θ represents the angle between the x-axis and a principal direction (x'-axis) of the medium, it also corresponds to a unique permeability matrix $[K]$ for a given anisotropic medium (see Equations 3.33). As described in Equations 3.32, the elements of $[K]$ matrix are functions of a single variable θ for a given anisotropic medium.
b. Since the permeability coordinates of point A are $K_{xx} = K'_{xx}$ and $K_{xy} = 0$, point A must correspond with $\theta = 0$ (Equations 3.33). Evidently, the generic point $P(K_{xx}, K_{xy})$ corresponds with parameter θ. It is also clear from Figure 3.9 that to make the generic point P approach point A, the

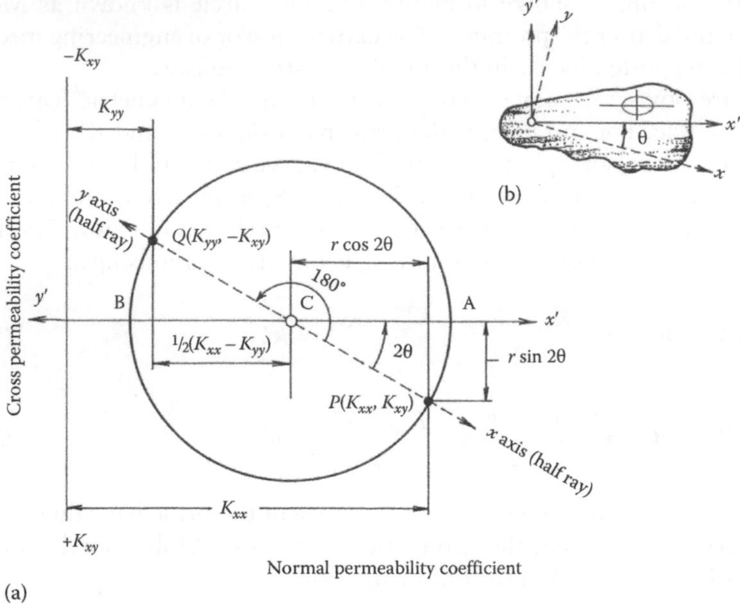

Figure 3.9 (a) Permeability plane and (b) physical plane.

radius *CP* must be rotated in a counterclockwise direction through an angle 2θ on the Mohr diagram. On the physical plane (Figure 3.9b), however, the *x*-axis must be rotated through an angle θ in a counterclockwise direction to align with the *x*′-axis. *It is to achieve the same sense of rotation in both the permeability plane and the physical plane that we chose the positive direction of* K_{xy}*-axis downward, contrary to the established convention of the analytical coordinate geometry.*

c. Finally, the radial directions *CP* and *CQ* in Figure 3.9 indeed represent the positive half rays of *x*-axis and *y*-axis, respectively.

3.4.2 Physical significance of cross permeability terms

Darcy's law for an anisotropic aquifer with respect to a rectangular, right-handed, Cartesian coordinate system is given as

$$\begin{Bmatrix} q_x \\ q_y \end{Bmatrix} = -\begin{bmatrix} K_{xx} & K_{xy} \\ K_{yx} & K_{yy} \end{bmatrix} \begin{Bmatrix} \dfrac{\partial \phi}{\partial x} \\ \dfrac{\partial \phi}{\partial y} \end{Bmatrix} \tag{3.22c}$$

where K_{xx} and K_{yy} are always positive coefficients (Equations 3.31). The cross permeability coefficients

$$K_{xy} = K_{yx} = \frac{1}{2}\left(K'_{xx} - K'_{yy}\right)\sin 2\theta \tag{3.42}$$

could, however, be either positive or negative depending on the value of θ and the sign of the term $\left(K'_{xx} - K'_{yy}\right)$. What is the physical meaning of the *negative sign* of the cross permeability coefficient? The answer to this question is presented in the following paragraphs.

For present discussion, we shall assume K'_{xx} is greater than K'_{yy}; in other words, θ denotes the angle between the x-axis and the major principal direction of permeability of the anisotropic medium (Figure 3.9b). Now, if θ is chosen to lie in the first quadrant, $0<\theta \leq \pi/2$, the sin 2θ will remain positive. Thus, in this case, as a consequence, the cross permeability coefficients will also remain positive, that is, $K_{xy}=K_{yx}\geq 0$. Let us now look at the corresponding flow field in the medium under the influence of a unit hydraulic gradient acting in the positive x-direction:

$$-\begin{Bmatrix} \dfrac{\partial \phi}{\partial x} \\ \dfrac{\partial \phi}{\partial y} \end{Bmatrix} = \begin{Bmatrix} 1 \\ 0 \end{Bmatrix} \tag{3.43}$$

By substituting Equation 3.43 in Equation 3.22c, we observe that

$$q_x = K_{xx} \geq 0 \tag{3.44a}$$

$$q_y = K_{yx} \geq 0 \tag{3.44b}$$

Thus, as long as the major principle axis of the anisotropic aquifer lies in the first quadrant of the x, y-coordinate *plane*, both components, q_x and q_y, of the specific discharge vector will remain positive under the influence of a hydraulic gradient vector acting in the positive x-direction This situation is graphically illustrated in Figure 3.10a. However, if the major principal axis is chosen to lie in the second quadrant of the x, y-plane, $(\pi/2 < \theta \leq \pi)$, the cross permeability coefficients become negative. Thus, in this case, the components of the specific discharge vector, under the influence of the hydraulic gradient vector acting in the positive x-direction, satisfy the following inequalities:

$$q_x = K_{xx} \geq 0 \tag{3.45a}$$

$$q_y = K_{yx} \leq 0 \tag{3.45b}$$

This situation is illustrated in Figure 3.10b.

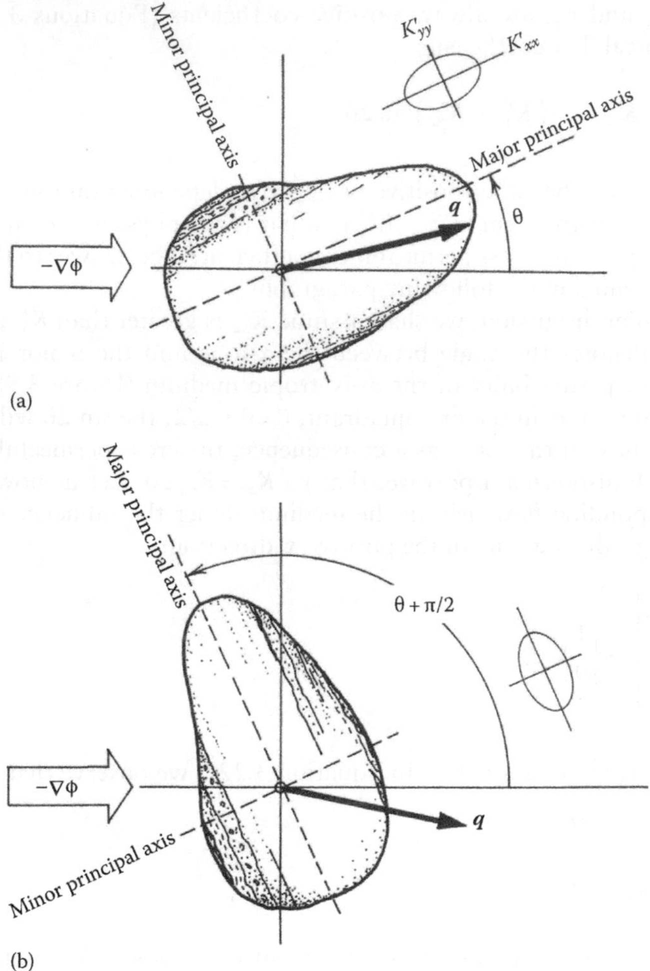

Figure 3.10 (a) Major principal direction of permeability lies in the first quadrant and (b) major principal direction of permeability lies in the second quadrant.

To grasp fully the significance of cross permeability coefficients in determining the direction of flow field, Figure 3.11 has been prepared. In this figure, the major principal direction of permeability is kept *constant to lie in the first quadrant*, while the influence of the hydraulic gradient on the direction of specific discharge vector q is graphically shown. The direction of the hydraulic gradient vector $-\nabla\phi$ is varied in three steps: starting from positive x-direction, to positive y-direction, to negative x-direction, and finally to negative y-direction. The corresponding directions of the specific discharge vector q are represented by

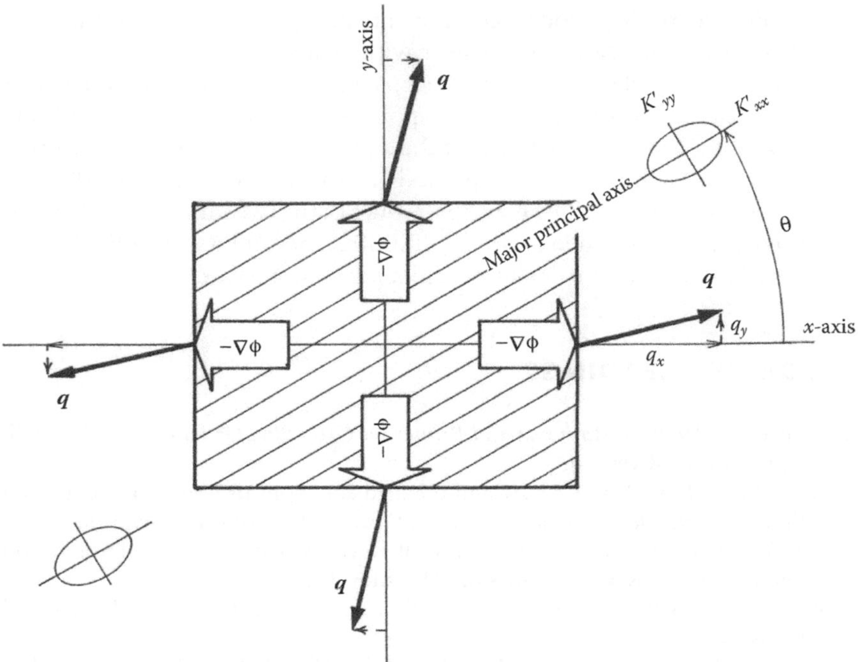

Figure 3.11 Role of cross permeability coefficients in tilting the specific discharge vector toward the major principal direction.

solid arrows in the figure. It is apparent from this figure that *the role of cross permeability coefficients is such that it always enforces the specific discharge vector to tilt toward the major principal direction of permeability*. The same conclusion can also be drawn if the principal direction of permeability of an anisotropic aquifer lies in the second quadrant of the x, y-coordinate *plane*.

3.5 EXERCISES

3.1 A variable head permeameter is shown in Figure 3.5. The permeameter of this kind is used in the laboratory to determine the coefficient of permeability of low-permeable soils. Find the coefficient of permeability of soil sample if $h_0 = 50$ cm, $a = 0.5$ cm^2, $A = 10$ cm^2, $L = 20$ cm and the observed head $h = 28$ cm after 1 h.

3.2 The permeability values of a two-dimensional, anisotropic, homogeneous aquifer are $K_{xx} = 0.100$ cm/s, $K_{yy} = 0.020$ cm/s, and $K_{xy} = K_{yx} = 0.003$ cm/s. Find (1) the directions and principal values of the permeability tensor and (2) the gradient of the piezometric surface at a point where the velocity components are $q_x = 0.10$ cm/s and $q_y = 0.1$ cm/s. Please illustrate your answer with neat sketches.

3.3 Considering a planar flow in the x, y-plane, find the components of the discharge vector (q_x, q_y) at a point where $d\phi/dx = 0.002$ and $d\phi/dy = 0.000$. You may assume that the aquifer is homogeneous, anisotropic aquifer with major principal axis inclined at an angle of 30° with respect to the x-axis. The principal values of the two-dimensional anisotropic permeability tensor are $K_1 = 0.0002$ cm/s and $K_2 = 0.00001$ cm/s. Please do not forget to mention units for your answer.

SUGGESTED READINGS

Hermance, J. F. 1999. *A Mathematical Primer on Groundwater Flow*, Prentice Hall, Upper Saddle River, NJ.

Hubbert, M. K. 1957. Darcy's Law and the field equations of flow of underground fluids, *Bulletin de l'Association Internationale d'Hydrologie Scientifique,* n° 5, 1957. Also, Publication No. 104, Shell Development Company, Exploration and Production Research Division, Houston, TX.

Strack, O. D. L. 1989. *Groundwater Mechanics*, Prentice Hall, Inc., Englewood Cliffs, NJ.

Todd, D. K. 1959. *Ground Water Hydrology*, John Wiley & Sons, Inc., New York.

Verruijt, A. 1970. *Theory of Groundwater Flow*, Macmillan and Co Ltd, London, U.K.

Wang, H. F. and M. P. Anderson. 1982. *Introduction to Groundwater Modeling Finite Difference and Finite Element Methods*, W. H. Freeman and Company, San Francisco, CA.

Chapter 4

Discharge potentials for two-dimensional flows in horizontal, shallow aquifers

4.1 HORIZONTAL, SHALLOW, CONFINED (ARTESIAN) AQUIFER

In general, the steady-state flow of an incompressible, homogeneous fluid in a three-dimensional, isotropic, homogeneous aquifer is governed by the Laplace equation

$$\left[\frac{\partial^2\phi}{\partial x^2}+\frac{\partial^2\phi}{\partial y^2}+\frac{\partial^2\phi}{\partial z^2}\right]\equiv\nabla^2\phi=0 \tag{3.20}$$

where ϕ represents the piezometric head (or level) at any point in the flow field. Of particular interest to us is the case when ϕ does not depend on one of the spatial coordinates, say z, where z-direction is taken vertically upward. In such a case, the piezometric head (or level) becomes a function of two independent variables, that is, $\phi=\phi(x, y)$, and the Laplace equation reduces to a two-dimensional case:

$$\left[\frac{\partial^2\phi}{\partial x^2}+\frac{\partial^2\phi}{\partial y^2}\right]\equiv\nabla^2\phi(x,y)=0 \tag{4.1}$$

Now, consider an aquifer bounded by two horizontal confining layers, some finite distance H apart (Figure 4.1). When H is much smaller than the areal extent, the aquifer is called *a horizontal, shallow, confined aquifer*. If we assume that the areal extent is infinite, $(-\infty<x<+\infty)$ and $(-\infty<y<+\infty)$, and there is no source or sink in the finite plane, then the following two inferences can be drawn:

1. *At a given (fixed) value of z, on any two verticals, say at (x_1,y_1) and (x_2,y_2), the specific discharge vectors q must be identical.* Borrowing the terminology from fluid mechanics, the velocity profiles at the two

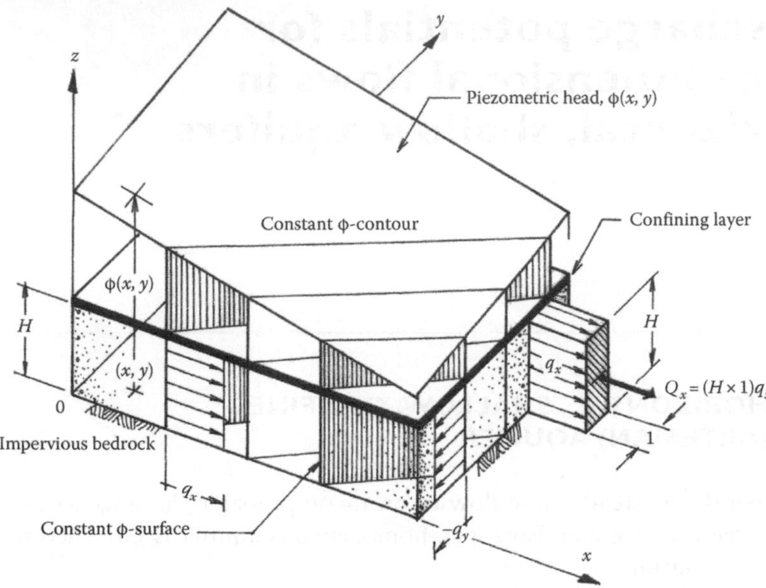

Figure 4.1 Definition sketch for horizontal, shallow, confined (artesian) aquifer.

verticals must be identical. This inference follows from our a priori intuition that both verticals can be considered as the centroidal axis for the infinite horizontal, confined aquifer.

2. *At any point in the flow field, the specific discharge vector **q** lies in a horizontal plane.* This inference follows from the previous assertion. The argument goes like this. Since the velocity profiles at any two verticals are identical, the partial derivatives of q with respect to x and y must be zero. Thus, the continuity equation (Equation 3.17) reduces to the following ordinary differential equation:

$$\frac{dq_z}{dz} = 0$$

whose solution is $q_z = C$, an arbitrary constant throughout the flow region. Since $q_z = 0$, at the impervious boundaries, it follows that $q_z \equiv 0$. Thus, the specific discharge vector lies everywhere in a horizontal plane.

Now, if the aquifer is isotropic, then according to Darcy's law, the following is true:

$$q_x = -K\frac{\partial \phi}{\partial x}; \quad q_y = -K\frac{\partial \phi}{\partial y} \tag{4.2}$$

Since in an *isotropic* aquifer the constant ϕ-*surfaces* are always normal to the specific discharge vector, it follows that such surfaces must be generated by vertical straight line generators, as shown in Figure 4.1. In other words, ϕ is indeed a function of two independent variables, x and y in a *horizontal, confined aquifer of infinite areal extent*. Since constant ϕ-*surfaces* are vertical, each component, q_x or q_y, remains uniform, along any vertical line. It is therefore possible to define the following equalities:

$$Q_x = Hq_x = -HK\frac{\partial\phi}{\partial x} = -\frac{\partial(HK\phi)}{\partial x} = -\frac{\partial(HK\phi + C)}{\partial x} = -\frac{\partial\Phi}{\partial x} \tag{4.3a}$$

$$Q_y = Hq_y = -HK\frac{\partial\phi}{\partial y} = -\frac{\partial(HK\phi)}{\partial y} = -\frac{\partial(HK\phi + C)}{\partial y} = -\frac{\partial\Phi}{\partial y} \tag{4.3b}$$

$$\Phi(x,y) \equiv HK\phi(x,y) + C \tag{4.3c}$$

where

C is an arbitrary constant

Q_x and Q_y are discharges per width normal to the x-axis and y-axis, respectively, as shown in Figure 4.1 (for clarity, only Q_x is shown)

In Equations 4.3a and 4.3b, the first equality follows from the uniformity of q_x and q_y with depth, the second from Darcy's law, and the rest from calculus. Thus, for a horizontal, confined flow of an infinite areal extent, there exists a potential function $\Phi = KH\phi + C$ whose partial derivatives with respect to x and y yield the discharges Q_x and Q_y per width normal to x- and y-axis, respectively, through the entire depth of aquifer. We shall therefore call $\Phi(x, y)$ the *discharge potential for horizontal, confined flows of infinite areal extent.*

It must however be emphasized that for the existence of discharge potential, the requirement of an infinite areal extent is not necessary. What is pertinent is the fact that the first equality in Equations 4.3a and 4.3b

$$Q_x - Hq_x$$

$$Q_y = Hq_y$$

must be true. This would be true whenever q_x and q_y are invariant with depth at a given vertical. This in turn implies that the constant ϕ-*surfaces* should be vertical for the existence of discharge potential. Thus, the discharge potential Φ is defined for flows through horizontal, confined aquifers even if the aquifers are of finite areal extent, *so long as the constant ϕ-surfaces are vertical prismatic surfaces generated by straight line generators.* This is most likely the case when H is much smaller than the areal extent of the aquifer—especially in the middle of the aquifer away from the boundary conditions. The discharge potential $\Phi(x, y)$ given in Equations 4.3 may therefore be applicable to flows through *horizontal, shallow, confined aquifers.*

The solutions of Laplace's equations are called the *harmonic functions.* It is one of the properties of the harmonic functions that if there are two harmonic functions, Φ_1 and Φ_2, then their linear combination

$$\Phi = \alpha\Phi_1 + \beta\Phi_2; \quad (\alpha, \beta \text{ denote arbitrary constants})$$

is also a harmonic function. Since the piezometric head $\phi(x, y)$ is a harmonic function (Equation 4.1), and any constant function is always a harmonic function, it follows from the preceding property that

$$\Phi(x,y) = KH\phi(x,y) + C \tag{4.4}$$

is also a harmonic function. Thus, the solution of a steady-state groundwater flow through a horizontal, shallow, confined aquifer is completely defined by the two-dimensional Laplace equation:

$$\left[\frac{\partial^2\Phi}{\partial x^2} + \frac{\partial^2\Phi}{\partial y^2}\right] \equiv \nabla^2\Phi(x,y) = 0 \tag{4.5}$$

in terms of the *discharge potential for horizontal, shallow, confined flow*, $\Phi = \Phi(x, y)$, subject to the appropriate boundary conditions. The discussion of discharge potential given here is similar to the treatment given by Strack (1989).

4.2 HORIZONTAL, SHALLOW, UNCONFINED (PHREATIC) AQUIFER

Figure 4.2 represents the essential physical features of a two-dimensional flow (in x, z-plane) in a phreatic aquifer. In this illustration, all sections parallel to x, z-plane are identical. While the bottom surface of flow is horizontal, the top surface is in general a curvilinear plane, which represents the watertable, or the phreatic surface, where the pressure is

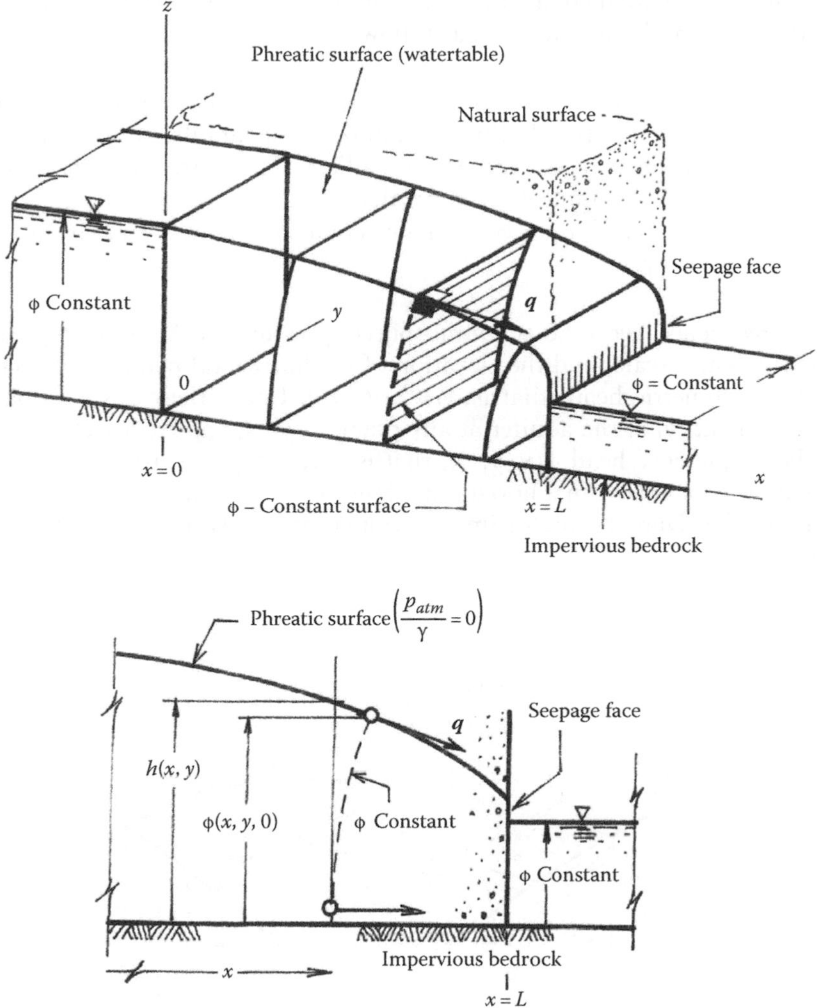

Figure 4.2 Physical features of a two-dimensional flow (in x, z-plane) in a phreatic aquifer.

atmospheric. At each vertical boundary, $x = 0$ or $x = L$, the piezometric head is a constant, because of the presence of a large body of water in a quasi-static state. Although the steady flow in an aquifer can be solved by finding the mathematical solution of the boundary-value problem associated with the Laplace equation, in this case, it is not possible because the flow region, where the Laplace equation applies, is not known a priori. This shortcoming has, however, been overcome by invoking the so-called

Dupuit (1863)–Forchheimer (1886) assumptions. The Dupuit–Forchheimer assumptions can be summarized as follows:

1. The constant ϕ-*surfaces* are assumed to be prismatic surfaces generated by vertical straight line generators. The hydraulic gradient at any vertical is taken as the (negative) gradient of the phreatic surface and it acts at every point of the vertical.
2. The specific discharge vector at a given (fixed) vertical remains invariant with depth.

For convenience, we take the atmospheric pressure as the zero, $p_{atm}=0$, of the pressure scale, and the elevation of the horizontal plane as the zero of the piezometric head, that is, $\phi=0$ *at* $z=0$. Under these assumptions, the depth $h(x, y)$ of the aquifer at any point (x, y) is, approximately, equal to the piezometric head $\phi(x, y, 0)$, that is $h(x, y)\cong\phi(x, y, 0)$. Thus, for a two-dimensional flow in a horizontal plane (x, y-plane) through a phreatic aquifer, the Dupuit–Forchheimer formulation leads to the following (Figure 4.3):

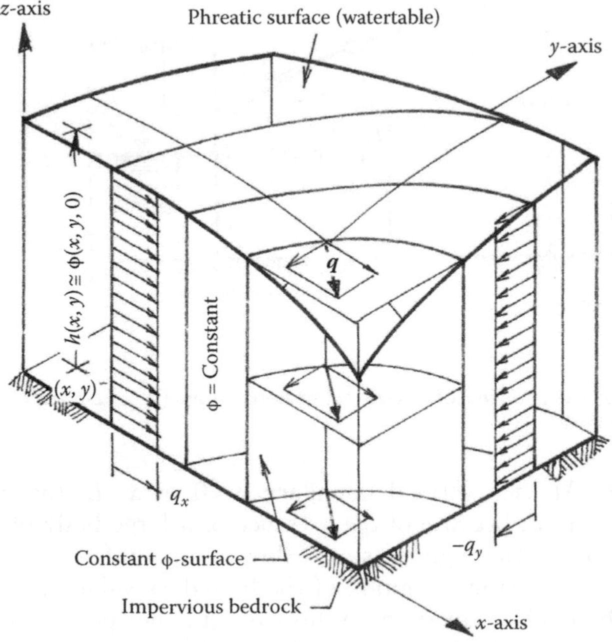

Figure 4.3 Two-dimensional flow (in x, y-plane) in a phreatic aquifer after the Dupuit–Forchheimer formulation.

$$Q_x = hq_x = \phi\left(-K\frac{\partial \phi}{\partial x}\right) = -\frac{\partial\left(\frac{K\phi^2}{2}+C\right)}{\partial x} = -\frac{\partial \Phi}{\partial x} \tag{4.6a}$$

$$Q_y = hq_y = \phi\left(-K\frac{\partial \phi}{\partial y}\right) = -\frac{\partial\left(\frac{K\phi^2}{2}+C\right)}{\partial y} = -\frac{\partial \Phi}{\partial y} \tag{4.6b}$$

$$\Phi(x,y) \equiv \frac{K\phi^2(x,y)}{2} + C \tag{4.6c}$$

In the preceding development, the approximation $\phi \cong h$ has been used, and C is the usual arbitrary constant. Since in the case of a phreatic aquifer the discharge potential Φ is nonlinearly related to the piezometric head ϕ, one naturally raises the question: Does Φ satisfy the Laplace equation? The short answer is yes! It can, however, be deduced by analyzing the continuity of flow in the differential element shown in Figure 4.4. If we set $\delta x = \delta y = 1$, it can be quickly observed from this figure that the flow satisfies the following:

$$\frac{\partial Q_x}{\partial x} + \frac{\partial Q_y}{\partial y} = 0 \tag{4.7}$$

Now, substituting the expressions for Q_x and Q_y from Equations 4.6 into Equation 4.7 yields

$$\left[\frac{\partial^2 \Phi}{\partial x^2} + \frac{\partial^2 \Phi}{\partial y^2}\right] \equiv \nabla^2 \Phi(x,y) = 0 \tag{4.8}$$

This shows that the discharge potential exists even for flows through a phreatic aquifer, and it satisfies the two-dimensional Laplace equation *under the Dupuit–Forchheimer assumptions.*

We can therefore conclude that for both *confined and unconfined* shallow horizontal flows, discharge potentials exist and they are harmonic functions with the following property:

$$Q_n = -\frac{\partial \Phi}{\partial n} \tag{4.9}$$

where

n denotes the spatial coordinate

Q_n denotes the steady-state discharge per width of aquifer normal to n-axis

The validity of the preceding equation can be demonstrated as follows.

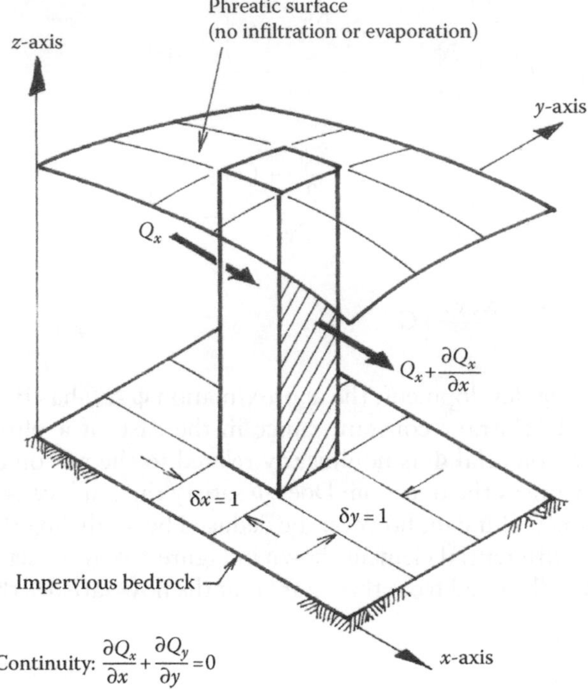

Figure 4.4 Continuity of flow in terms of discharges Q_x and Q_y.

Let the discharges Q_x and Q_y be defined with respect to the x, y-coordinate system and let the angle between the x-axis and the n-axis be θ, as shown in Figure 4.5. Now, the partial derivatives of $\Phi(x, y)$ are related by the following equation (Equation 3.26):

$$\frac{\partial \Phi}{\partial n} = \cos\theta \frac{\partial \Phi}{\partial x} + \sin\theta \frac{\partial \Phi}{\partial y} \tag{4.10a}$$

or, after substituting the values for $\cos\theta$ and $\sin\theta$ from Figure 4.5, the preceding equation reduces to

$$\frac{\partial \Phi}{\partial n} = \frac{\Delta y}{\Delta s}\frac{\partial \Phi}{\partial x} + \frac{\Delta x}{\Delta s}\frac{\partial \Phi}{\partial y} \tag{4.10b}$$

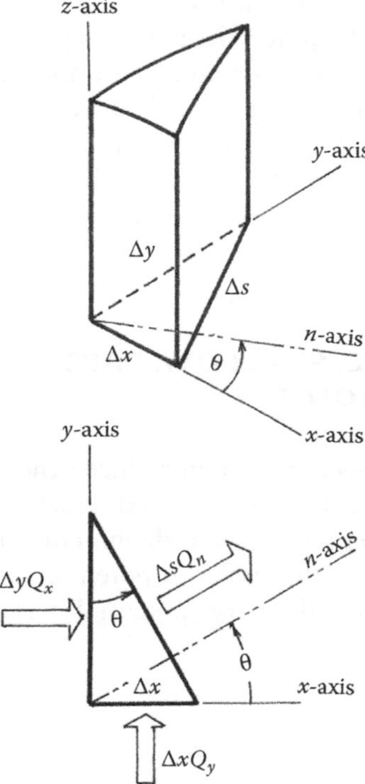

Figure 4.5 Definition sketch for n-axis and Q_n.

or

$$\Delta s \frac{\partial \Phi}{\partial n} = \Delta y \frac{\partial \Phi}{\partial x} + \Delta x \frac{\partial \Phi}{\partial y} \tag{4.10c}$$

$$\Delta s Q_n = \Delta y\, Q_x + \Delta x\, Q_y \tag{4.10d}$$

The last equation is simply the reaffirmation of the fact that Q_n satisfies the continuity requirement. Thus,

$$Q_n = -\frac{\partial \Phi}{\partial n}$$

indeed represents the discharge per width normal to the n-axis.

In summary, we reiterate that in the general case of *planar flows in x, y-plane*, the discharge potential $\Phi(x, y)$ exists, irrespective of whether the flow takes place through a confined aquifer or an unconfined aquifer. In the case of an unconfined aquifer, the flow is represented as a planar flow field by invoking the Dupuit–Forchheimer assumptions. Thus, in both cases, the flow field is assumed as the planar flow field in x, y-plane. Furthermore, the discharge potential in both cases is a harmonic function with the property that $Q_x = -\partial\Phi/\partial x$ or $Q_y = -\partial\Phi/\partial y$, where Q_x or Q_y represents the discharge per width of aquifer normal to the x-axis or the y-axis, respectively, through the entire depth of aquifer.

4.3 HORIZONTAL, SHALLOW, PARTLY CONFINED AQUIFER

In the previous two sections, we introduced the notion of discharge potentials for a horizontal, shallow, confined aquifer and for a shallow, unconfined (phreatic) aquifer resting on horizontal impervious bedrock. It was also noted that in both cases, the discharge potential satisfies the two-dimensional Laplace equation and it (discharge potential) has the following property:

$$Q_n = -\frac{\partial\Phi}{\partial n} \tag{4.9}$$

where

n denotes the spatial coordinate

Q_n denotes the discharge per width of aquifer normal to n-axis

In many practical problems, we are encountered with the situation where the flow takes place in a confined aquifer, in part of the flow region, and in an unconfined (phreatic) aquifer, in the remaining part of the aquifer. Two typical examples are illustrated in Figure 4.6. Following Strack (1989), we shall designate the boundary between the two regions as the interzonal boundary.

We know that the piezometric head, $\phi = z + p/\gamma$, is defined at each point of the flow region, irrespective of whether the point lies in the confined or unconfined region. If the pressure field is continuous throughout the flow region, it follows that ϕ is also continuous throughout the flow region—especially across the interzonal boundary. Furthermore, since ϕ is defined with respect to an arbitrary datum, we can the datum at the horizontal impervious base as shown in Figure 4.6.

Now, let us look at the discharge potentials. For horizontal, shallow confined aquifers, it is defined as

$$\Phi_c(x,y) = KH\phi(x,y) + C_c \tag{4.11}$$

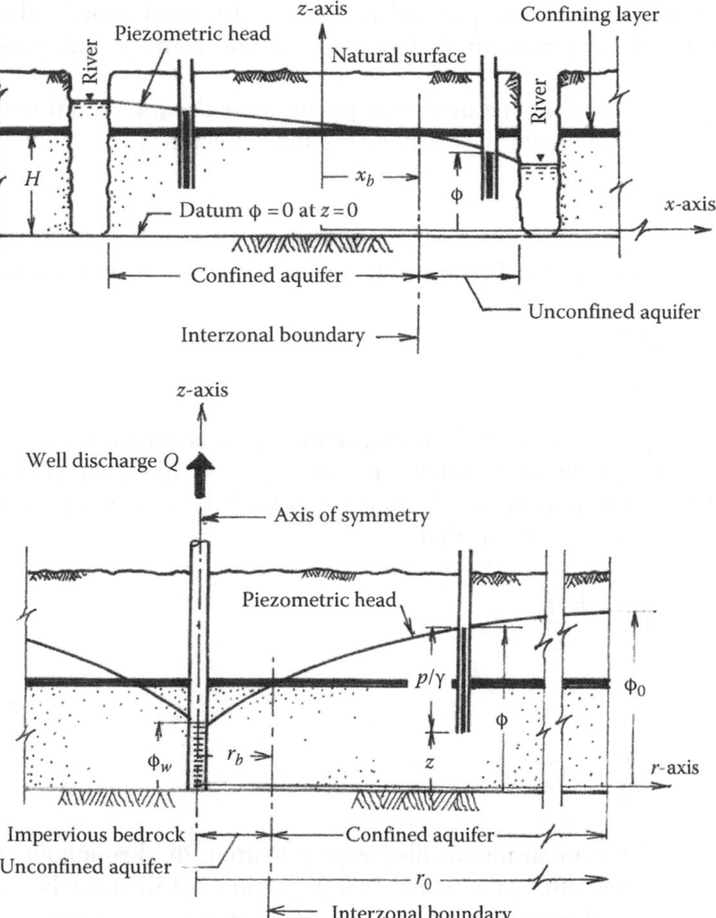

Figure 4.6 Examples of one-dimensional flows in partly confined aquifers.

and for horizontal, shallow, unconfined aquifer (under Dupuit–Forchheimer's assumptions) as

$$\Phi_u(x,y) = \frac{K\phi^2(x,y)}{2} + C_u \tag{4.12}$$

In the previous equations, the subscripts c and u refer to confined and unconfined regions of the aquifer, respectively. It is also understood that discharge potentials are defined when constant φ-*surfaces* represent either actual, or assumed, prismatic surfaces generated by vertical straight line generators (parallel to z-axis). In other words, discharge potential only depends on x- and y-coordinates, and is independent of z-coordinate.

Now, let us look at the discharge potential at the interzonal boundary. When approached from the confined region, it becomes

$$\Phi_c = KHH + C_c \tag{4.13}$$

Likewise, when approached from the unconfined region, it becomes

$$\Phi_u = \frac{KH^2}{2} + C_u \tag{4.14}$$

because piezometric head, ϕ, is continuous as a consequence of continuity of pressure field. If we further require that the discharge potential, Φ, should be continuous across the interzonal boundary, then it follows from the previous two equations that

$$KH^2 + C_c = \frac{KH^2}{2} + C_u \tag{4.15a}$$

or

$$C_c = -\frac{KH^2}{2} + C_u \tag{4.15b}$$

In order to have a continuous discharge potential, Φ, throughout the flow region, it is, therefore, necessary that Equation 4.15b must be satisfied. This in turn implies that only one of the two arbitrary constants, C_c or C_u, can be left arbitrary. We, therefore, set arbitrarily C_u to be zero. Thus, the definition Equations 4.11 and 4.12 reduce to

$$\Phi_c(x,y) = KH\phi(x,y) - \frac{KH^2}{2} \tag{4.16}$$

$$\Phi_u(x,y) = \frac{K\phi^2(x,y)}{2} \tag{4.17}$$

for confined and unconfined regions of flow, respectively.

Finally, since at the interzonal boundary the discharge potential is constant ($\Phi_c = \Phi_u = 1/2KH^2$), there can be no flow tangential to the boundary. Hence, discharge is continuous across the interzonal boundary. In other words,

$$Q_n = -\frac{\partial \Phi_c}{\partial n} = -\frac{\partial \Phi_u}{\partial n} \tag{4.18}$$

where

n-axis is normal to the interzonal boundary and it points in the flow direction

Q_n represents the discharge, per width normal to n-axis through the entire depth H of aquifer, across the interzonal boundary

Thus, we conclude that the discharge potential and its partial derivatives are continuous across the interzonal boundary.

4.4 APPLICATIONS

In this section, we apply the notion of discharge potential to solve illustrative problems related to steady groundwater flow in three types of horizontal, shallow, aquifers: (1) confined, (2) unconfined, and (3) partly confined aquifers. In all these types, discharge potential Φ exists and it satisfies the Laplace equation in two dimensions with appropriate boundary conditions.

4.4.1 Flow through horizontal, shallow, confined (artesian) aquifer

4.4.1.1 Case I: Rectilinear flow through a confined aquifer

As an elementary illustration, let us consider a shallow ($H \ll L$) confined aquifer completely traversed by two parallel perennial rivers extending from negative infinity to positive infinity as shown in Figure 4.7. Let the two rivers be at a distance L apart, and at a typical cross section, the water depths in the rivers be denoted by h_1 and h_2. We shall further assume that the water surface slopes in the two perennial rivers are negligible in comparison with $(h_1 - h_2)/L$. This is tantamount to saying that water surface elevation in each river does not change with y. Thus, all sections parallel to x, z-plane are identical and the discharge potential does not depend on y.

Under these simplifying assumptions, the specific discharge can be assumed rectilinear in x-direction. Thus, the two-dimensional Laplace equation given in Equation 4.5 reduces to the following ordinary differential equation:

$$\frac{d^2\Phi}{dx^2} = 0 \tag{4.19}$$

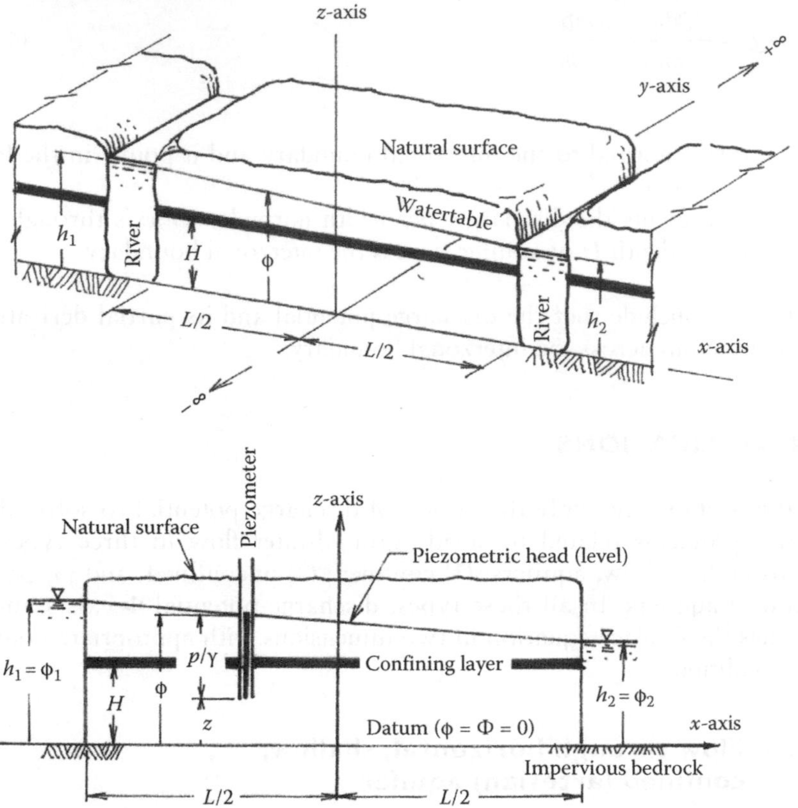

Figure 4.7 **Definition sketch for flow through confined (artesian) aquifer.**

because $\partial\Phi/\partial y \equiv 0$. The general solution of the preceding differential equation is

$$\Phi = Ax + B \tag{4.20}$$

where A and B are arbitrary constants. These constants can be determined by invoking the boundary conditions. However, before determining these constants, it is convenient—though not necessary—to assume the datum for ϕ and Φ to lie on the horizontal plane $z=0$, that is, at $z=0$, $\phi=0$, and $\Phi=0$. Thus, at the two rivers, the piezometric head satisfies $\phi_1=h_1$ and $\phi_2=h_2$, as shown in the figure. This yields the following boundary conditions at the left and right boundaries, respectively:

$$\text{At } x=\frac{-L}{2}, \quad \phi=\phi_1=h_1, \quad \text{or} \quad \Phi=\Phi_1=KH\phi_1=KHh_1 \tag{4.21a}$$

$$\text{At } x = \frac{L}{2}, \quad \phi = \phi_2 = h_2, \quad \text{or} \quad \Phi = \Phi_2 = KH\phi_2 = KHh_2 \tag{4.21b}$$

Substituting these boundary conditions in Equation 4.20 yields

$$\Phi_1 = A\left(\frac{-L}{2}\right) + B \tag{4.22a}$$

$$\Phi_2 = A\left(\frac{+L}{2}\right) + B \tag{4.22b}$$

or

$$A = \frac{-(\Phi_1 - \Phi_2)}{L} \tag{4.23a}$$

$$B = \frac{(\Phi_1 + \Phi_2)}{2} \tag{4.23b}$$

Hence, the discharge potential for this case becomes

$$\Phi = \frac{-(\Phi_1 - \Phi_2)}{L} x + \frac{(\Phi_1 + \Phi_2)}{2} \tag{4.24}$$

This finally yields the discharge per width normal to the x-axis through the entire depth of the aquifer as

$$Q = Q_x = -\frac{d\Phi}{dx} = \frac{(\Phi_1 - \Phi_2)}{L} \tag{4.25}$$

Equation 4.24 can also be written in terms of the piezometric head (or level) as

$$\phi(x) = \frac{-(\phi_1 - \phi_2)}{L} x + \frac{(\phi_1 + \phi_2)}{2} \tag{4.26}$$

which gives the specific discharge as

$$q = -K\frac{d\phi}{dx} = \frac{K(\phi_1 - \phi_2)}{L} = \frac{K(h_1 - h_2)}{L} \tag{4.27}$$

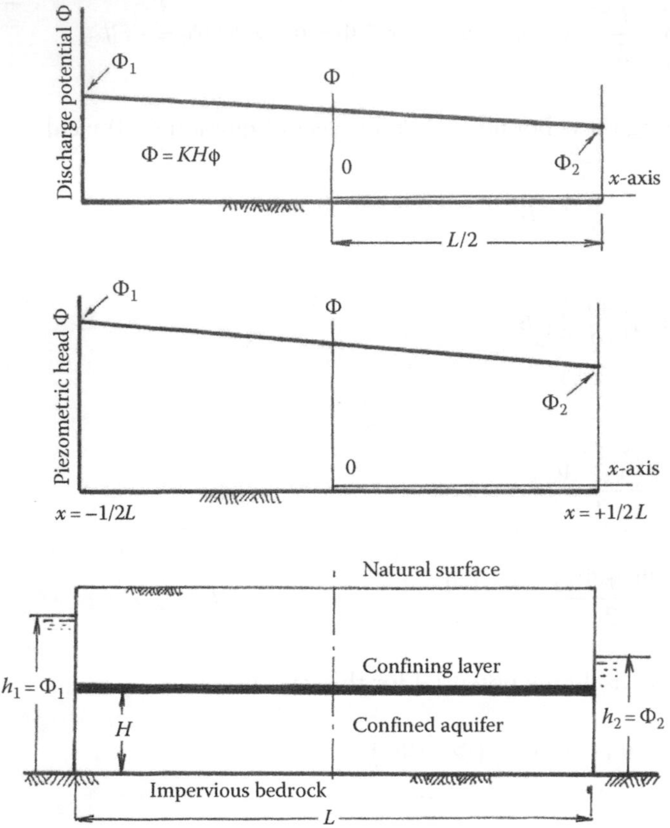

Figure 4.8 Variation of discharge potential and piezometric head.

In this case, Equations 4.24 and 4.26 show that both the discharge potential Φ and the piezometric head ϕ vary linearly with x. This fact is graphically shown in Figure 4.8.

4.4.1.2 Case II: Radial flow toward a well in a confined (artesian) aquifer

As a second example, we shall consider the radial flow toward a well centrally located in a circular island. In this case, we are interested in the steady well discharge, Q, and its relationship with the discharge potential, Φ. For further analysis, it is more convenient to use a cylindrical coordinate system (r, θ, z). In horizontal, shallow confined aquifers, the discharge potential does not depend on z-coordinate. Furthermore, because of axial symmetry, the discharge potential only depends on r and not on θ, that is, $\Phi = \Phi(r)$. Thus,

$$Q_r = -\frac{\partial \Phi}{\partial r} = -\frac{d\Phi}{dr} \tag{4.28}$$

and the well discharge Q is obtained from the requirement of continuity of flow (see Figure 4.9):

$$Q = (2\pi r)(-Q_r) = 2\pi r \frac{d\Phi}{dr} \tag{4.29}$$

Rearranging the preceding equation yields the following ordinary differential equation that governs the discharge potential:

$$d\Phi = \frac{Q}{2\pi}\frac{dr}{r} \tag{4.30}$$

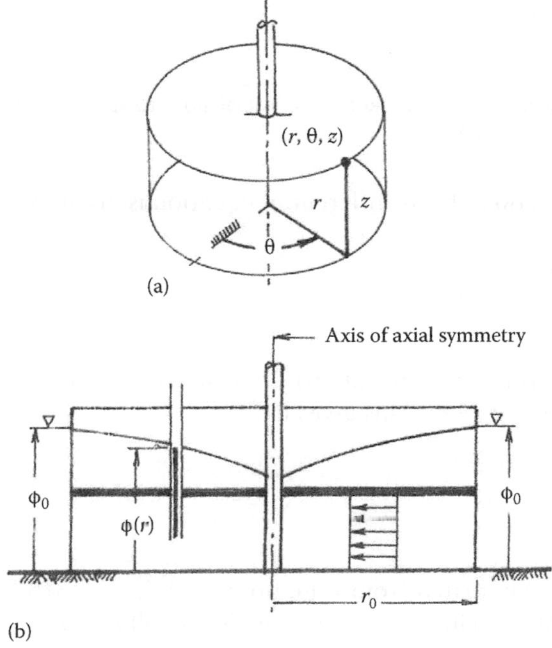

Figure 4.9 Definition sketch for radial flow toward a well: (a) cylindrical coordinate system, (b) symmetrical cross section. *(Continued)*

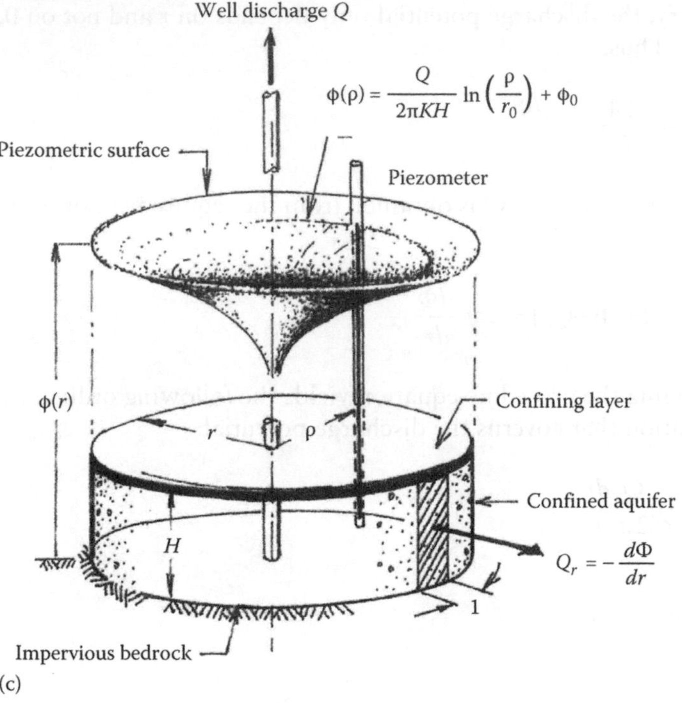

Figure 4.9 (Continued) Definition sketch for radial flow toward a well: (c) perspective view.

The general solution of this differential equation is given as follows:

$$\Phi(r) = \frac{Q}{2\pi}\ln r + C \tag{4.31a}$$

where C is an arbitrary constant. To determine C, we invoke the boundary condition: $\Phi = \Phi_0$ at $r = r_0$. This gives us

$$\Phi_0 = \frac{Q}{2\pi}\ln r_0 + C \tag{4.31b}$$

Now C can be eliminated from Equations 4.31a and 4.31b by subtracting one from the other. This subtraction yields the following:

$$\Phi(r) = \frac{Q}{2\pi}\ln\frac{r}{r_0} + \Phi_0 \tag{4.32}$$

The previous equation can also be expressed in terms of the piezometric head ϕ by making use of the identity $\Phi \equiv KH\phi$. The substitution for Φ in Equation 4.32 yields

$$\phi(r) = \frac{Q}{2\pi KH} \ln \frac{r}{r_0} + \phi_0 \tag{4.33}$$

The preceding result was previously reported by Thiem in 1906.

4.4.2 Flow through horizontal, shallow, unconfined (phreatic) aquifer

4.4.2.1 *Case I: One-dimensional flow through an unconfined (phreatic) aquifer*

Let us consider a shallow phreatic aquifer, resting on horizontal impervious bedrock and completely traversed by two parallel perennial rivers L distance apart, extending from negative infinity to positive infinity, as shown in Figure 4.10. As with the first example (Case I) discussed previously, we shall assume that the slope of water surface in each perennial river is negligible. Thus, all cross sections, parallel to x, z-plane, are identical and the two-dimensional Laplace equation (4.8) reduces to the following ordinary differential equation

$$\frac{d^2\Phi}{dx^2} = 0 \tag{4.34}$$

because $\partial\Phi/\partial y \equiv 0$. If we assume, for convenience, that the piezometric head ϕ and the discharge potential Φ are measured with respect to the horizontal impervious bedrock, that is, $\phi = \Phi = 0$, at $z = 0$, the boundary conditions on the left and right side of the idealized flow region become, respectively,

$$\text{At } x = \frac{-L}{2}, \quad \phi = \phi_1 = h_1; \quad \text{or} \quad \Phi = \Phi_1 = \frac{K\phi_1^2}{2} = \frac{Kh_1^2}{2} \tag{4.35a}$$

$$\text{At } x = \frac{+L}{2}, \quad \phi = \phi_2 = h_2; \quad \text{or} \quad \Phi = \Phi_2 = \frac{K\phi_2^2}{2} = \frac{Kh_2^2}{2} \tag{4.35b}$$

The solution of the differential equation (4.34) subject to the boundary conditions (4.35a) and (4.35b) is exactly the same as given previously, that is,

$$\Phi = \frac{-(\Phi_1 - \Phi_2)}{L} x + \frac{(\Phi_1 + \Phi_2)}{2} \tag{4.36}$$

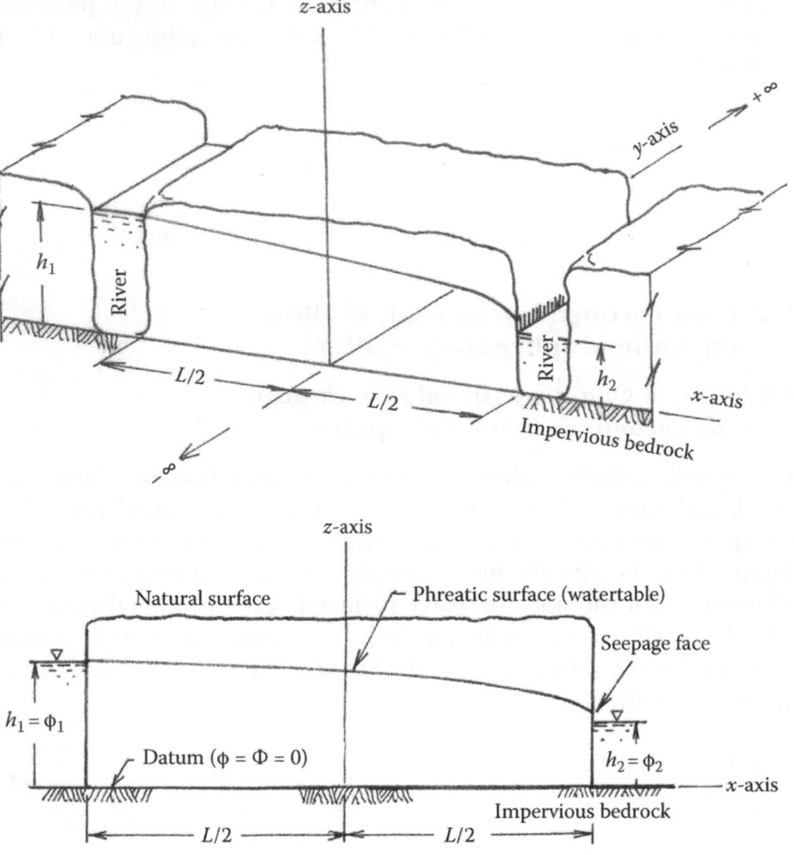

Figure 4.10 Definition sketch for flow through an unconfined (phreatic) aquifer.

This shows that the discharge potential, Φ, varies linearly with x for unidirectional flows even in the case of a *phreatic aquifer under the Dupuit–Forchheimer assumptions*. On the other hand, it should be emphasized that the piezometric head, ϕ, does not vary as a linear function of x, because Φ and ϕ are not related linearly in the case of phreatic aquifer. Recalling the definition of discharge potential for horizontal, shallow unconfined aquifer

$$\Phi \equiv \frac{K\phi^2}{2} + C \tag{4.6c}$$

we see that Φ is a quadratic function of ϕ. Furthermore, the constant C in the preceding equation vanishes due to the fact that the arbitrary datum for each function, ϕ or Φ, is set at $z=0$. Thus, for one-dimensional flow through the phreatic aquifer, Equation 4.6c reduces to

$$\Phi \equiv \frac{K\phi^2}{2} \tag{4.37}$$

Now, combining Equation 4.37 with Equation 4.36 yields

$$\phi^2 = -\frac{\phi_1^2 - \phi_2^2}{L}x + \frac{\phi_1^2 + \phi_2^2}{2} \tag{4.38a}$$

$$\phi^2(x) \cong h^2(x) = -\frac{h_1^2 - h_2^2}{L}x + \frac{h_1^2 + h_2^2}{2} \tag{4.38b}$$

where $h=h(x)$ represents the height of the phreatic surface above $z=0$ plane. Equation 4.38b represents the so-called Dupuit parabola, depicting the shape of the watertable (phreatic surface). The last equation can be written in a normalized form as indicated in the following text:

$$\frac{h(x)-h_2}{h_1-h_2} = \frac{\left(\left[\left\{(h_2/h_1)^2 - 1\right\}\frac{x}{L} + \frac{1+(h_2/h_1)^2}{2}\right]^{1/2} - h_2/h_1\right)}{(1-h_2/h_1)} \tag{4.38c}$$

Equation 4.38c is shown graphically in Figure 4.11 (lower plot) for selected values of head ratios, h_2/h_1. In this figure, the ordinate represents the normalized height, above h_2, of the watertable; the abscissa, the normalized x-coordinate in the direction of flow; and the curve parameter, the head ratio, h_2/h_1. When the head ratio $h_2/h_1=0$, the watertable profile is represented by a parabola whose vertex lies at point (0.5, 0), as shown in the figure. It is also apparent from this figure that the watertable profiles become more and more linear as h_2 approaches h_1. The upper plot shows the linear variation of the discharge potential from Φ_1 to Φ_2. It is apparent from these plots that while the watertable profile depends on the head ratio h_2/h_1, the variation of the discharge potential does not depend on this ratio—it varies linearly from Φ_1 to Φ_2 for all values of head ratios h_2/h_1.

Finally, the discharge in the x-direction per width normal to the x-axis through the entire depth of the aquifer is obtained by differentiating Equation 4.36 with respect to x. Thus,

$$Q = Q_x = -\frac{d\Phi}{dx} = \frac{(\Phi_1 - \Phi_2)}{L} \tag{4.39}$$

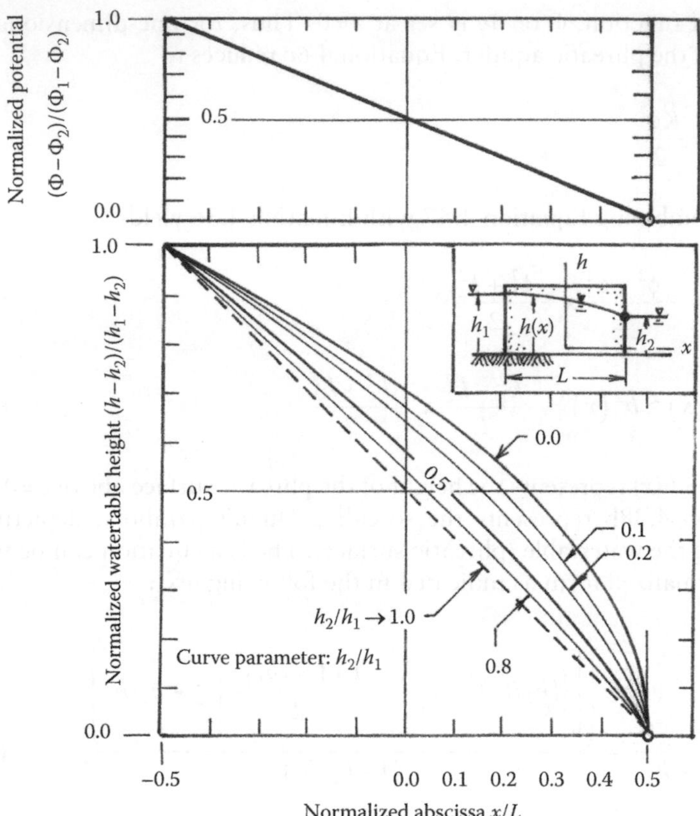

Figure 4.11 Normalized plot of the Dupuit parabola.

We can substitute the values

$$\Phi_1 = \frac{Kh_1^2}{2} \tag{4.40a}$$

$$\Phi_2 = \frac{Kh_2^2}{2} \tag{4.40b}$$

into Equation 4.39 to obtain

$$Q = K\left(\frac{h_1^2 - h_2^2}{2L}\right) \tag{4.41a}$$

This equation is also known as the Dupuit formula for discharge per width through the unconfined (phreatic) aquifer. It is, however, more instructive to rewrite Equation 4.41a in the following form:

$$Q = \left\{\frac{(h_1 + h_2)}{2}\right\}\left[K\left(\frac{h_1 - h_2}{L}\right)\right]$$

$$= \{\text{Average flow depth}\}[\text{Average specific discharge}] \tag{4.41b}$$

It is now clear that the Dupuit formula obtained by integrating the differential equation (4.34) using the Dupuit–Forchheimer formulation leads to the result that the discharge per width of the phreatic aquifer is simply the product of the average depth of flow with the average specific discharge. It is also of no surprise that this discharge is constant, because of one-dimensional nature of flow.

4.4.2.2 *Case II: Radial flow toward a well in an unconfined (phreatic) aquifer*

This case is similar to the one dealt with in the case of the confined aquifer. In terms of discharge potential Φ, the governing equation and the boundary conditions are identical to those given before. The only difference, however, is in the definition of the potential functions for confined and unconfined aquifers. Thus, Equation 4.32, obtained previously in connection with the well in a confined aquifer, can be used here:

$$\Phi = \frac{Q}{2\pi}\ln\frac{r}{r_0} + \Phi_0 \tag{4.32}$$

where Φ and Φ_0 are now defined, respectively, by the following equations for the unconfined aquifer:

$$\Phi = \frac{1}{2}K\phi^2 + C_u \tag{4.42a}$$

$$\Phi_0 = \frac{1}{2}K\phi_0^2 + C_u \tag{4.42b}$$

Substituting these values in Equation 4.32 yields the following:

$$\phi = \left(\frac{Q}{\pi K}\ln\frac{r}{r_0} + \phi_0^2\right)^{1/2} \tag{4.43}$$

A few concluding remarks are in order. Whereas the discharge potential Φ satisfies the two-dimensional Laplace equation for the unconfined aquifer, the piezometric head ϕ does not. Thus, superposition of Φ is allowed but that of ϕ is not. Furthermore, these analyses of radial flows to wells in confined as well as unconfined aquifers assume a constant head boundary at a finite radius r_0. The notion of a steady-state condition (or equilibrium) in an extensive aquifer where $r_0 \to \infty$ needs further discussion. This aspect of the well flow will be dealt with later.

4.4.3 Flow through horizontal, shallow, partly unconfined aquifer

4.4.3.1 *Case I: One-dimensional flow through partly unconfined aquifer*

A typical case of one-dimensional flow through a partly unconfined aquifer is shown in Figure 4.12. It may be reiterated that the discharge potential Φ satisfies the two-dimensional Laplace equation in the entire flow region, including confined as well as unconfined aquifers. Furthermore, we forced the discharge potential to be continuous everywhere within the flow region by imposing restrictions on the arbitrary constants (Section 4.3). The *continuous discharge potential* is given by the following definition equations:

$$\Phi(x,y) = \Phi_c(x,y) = KH\phi(x,y) - \frac{KH^2}{2} \quad \text{(for confined aquifer)} \tag{4.16}$$

$$\Phi(x,y) = \Phi_u(x,y) = \frac{K\phi^2(x,y)}{2} \quad \text{(for unconfined aquifer)} \tag{4.17}$$

From a mathematical point of view, the flow is again governed by the ordinary differential equation

$$\frac{d^2\Phi}{dx^2} = 0$$

subject to the boundary conditions: $\Phi = \Phi_1$ and $\Phi = \Phi_2$ at $x = -1/2\,L$ and $x = +1/2\,L$, respectively. The solution of this boundary-value problem is

$$\Phi = \frac{-(\Phi_1 - \Phi_2)}{L}x + \frac{(\Phi_1 + \Phi_2)}{2} \tag{4.44a}$$

in which Φ_1 and Φ_2 are defined by definition equations (4.16) and (4.17), respectively. Substituting Φ_1 and Φ_2 in Equation 4.44a yields

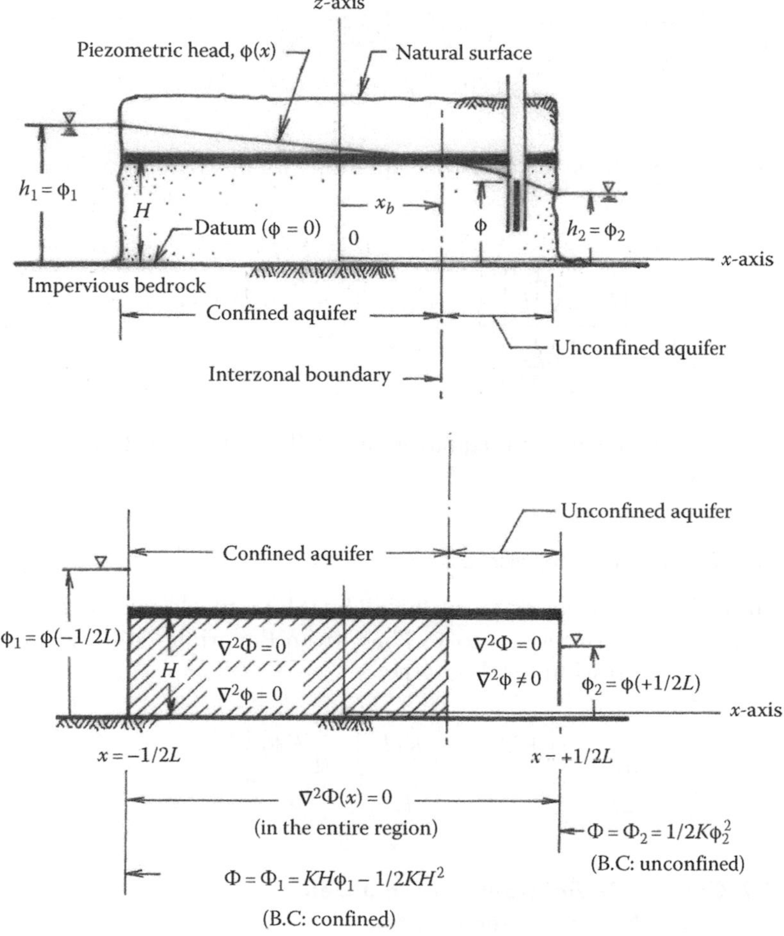

Figure 4.12 One-dimensional flow through partly unconfined aquifer.

$$\Phi = \frac{-\left[\left(KH\phi_1 - \frac{1}{2}KH^2\right) - \frac{1}{2}K\phi_2^2\right]}{L}x + \frac{\left(KH\phi_1 - \frac{1}{2}KH^2\right) + \frac{1}{2}K\phi_2^2}{2} \tag{4.44b}$$

4.4.3.1.1 Location of interzonal boundary

Let at the interzonal boundary $x = x_b$, the potential $\Phi = 1/2\,KH^2$. Substituting these values in Equation 4.44b yields the following expression:

$$\frac{x_b}{L} = \frac{\frac{1}{2}\left(KH^2 - KH\phi_1 + \frac{1}{2}KH^2 - \frac{1}{2}K\phi_2^2\right)}{-KH\phi_1 + \frac{1}{2}KH^2 + \frac{1}{2}K\phi_2^2} \tag{4.45a}$$

The right-hand side of the previous equation can be written in dimensionless form as follows:

$$\frac{x_b}{L} = \frac{\frac{1}{2}\left[\frac{3}{2} - \frac{\phi_1}{H} - \frac{1}{2}\left(\frac{\phi_2}{H}\right)^2\right]}{-\frac{\phi_1}{H} + \frac{1}{2} + \frac{1}{2}\left(\frac{\phi_2}{H}\right)^2} \tag{4.45b}$$

It is evident from the previous equation that $x_b/L \rightarrow -1/2$ or $1/2$ as $\phi_1/H \rightarrow 1.0$ or $\phi_2/H \rightarrow 1.0$.

4.4.3.1.2 *Determination of discharge*

As before, the discharge can be evaluated by taking the derivative of the discharge potential with respect to x. Thus, from Equation 4.44b, we obtain the following expression for discharge:

$$Q = Q_x = -\frac{d\Phi}{dx} = \frac{\left[\left(KH\phi_1 - \frac{1}{2}KH^2\right) - \frac{1}{2}K\phi_2^2\right]}{L} \tag{4.46}$$

4.4.3.2 Case II: Radial flow toward a well in partly unconfined aquifer

A typical case is illustrated in Figure 4.13. It is assumed that the aquifer is confined at a radial distance r_0, and unconfined at the well radius, r_w. The equation describing the (continuous) discharge potential Φ as a function of radial distance r is the same as given before for the case of completely confined, or completely unconfined aquifer, that is,

$$\Phi = \frac{Q}{2\pi}\ln\frac{r}{r_0} + \Phi_0 \tag{4.32}$$

Since at r_0 the aquifer is confined, the discharge potential is given by the following equation:

$$\Phi_0 = KH\phi_0 - \frac{1}{2}KH^2 \tag{4.47}$$

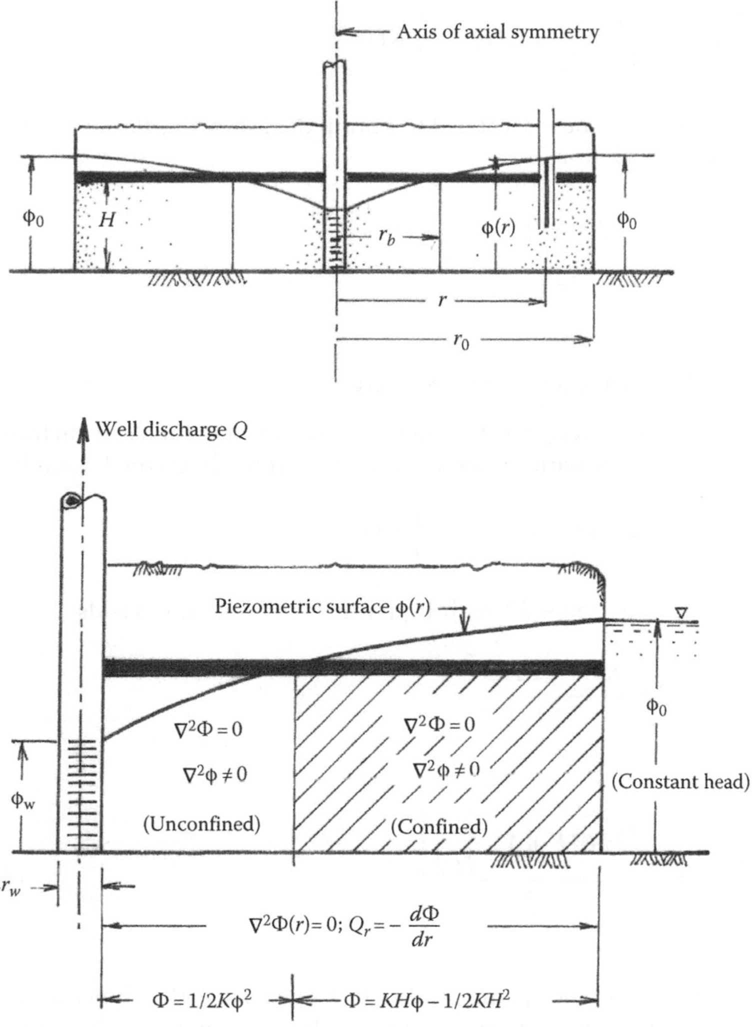

Figure 4.13 Radial flow through a partly unconfined aquifer.

Also, since the aquifer behaves as an unconfined aquifer at the well, the discharge potential at the well is given by the following:

$$\Phi_w = \frac{1}{2} K \phi_w^2 \tag{4.48}$$

where the subscript w refers to quantities at the well radius, r_w. Combining the previous three equations yields

$$\frac{1}{2}K\phi_w^2 = \frac{Q}{2\pi}\ln\frac{r_w}{r_0} + KH\phi_0 - \frac{1}{2}KH^2 \tag{4.49}$$

Equation 4.49 can be rearranged to obtain the following explicit expression for the well discharge:

$$Q = \frac{2\pi\left[\frac{1}{2}K\phi_w^2 - KH\phi_0 + \frac{1}{2}KH^2\right]}{\ln r_w/r_0} \tag{4.50}$$

4.4.3.2.1 Location of interzonal boundary

In this case, the interzonal boundary represents a cylindrical surface with the axis coincident with the axis of the well. At the interzonal boundary, let

$$r = r_b; \quad \phi = H; \quad \text{or} \quad \Phi = \frac{1}{2}KH^2 \tag{4.51}$$

Combining Equation 4.32 with Equations 4.47 and 4.51 yields

$$\ln\frac{r_b}{r_0} = 2\pi\frac{\left[KH^2 - KH\phi_0\right]}{Q} \tag{4.52a}$$

or

$$\frac{r_b}{r_0} = \exp\left[\frac{\left(2\pi KH^2\right)\left(1 - \frac{\phi_0}{H}\right)}{Q}\right] \tag{4.52b}$$

It is obvious that the quantity ($2\pi KH^2$) must have the dimension of Q, because the argument of an exponential function must be dimensionless. Let this quantity be denoted by Q_H so that $Q_H \equiv (2\pi KH^2)$. It is possible to assign a physical meaning to Q_H. For instance, it represents the (imaginary) radial discharge entering the aquifer cylinder of radius and height equal to H, under a unit hydraulic gradient. Using Q_H, the preceding equation can be written as

$$\frac{r_b}{r_0} = \exp\left[-\left(\frac{Q_H}{Q}\right)\left(\frac{\phi_0}{H} - 1\right)\right]; \quad \phi_0 \geq H \tag{4.52c}$$

in which quantities within the parentheses are positive. Thus, from this equation, one readily concludes that $r_b \to r_0$ as $\phi_0 \to H$.

4.5 ILLUSTRATIVE PROBLEMS

4.1 A typical cross section through a confined aquifer traversed by two parallel perennial rivers is shown in the following sketch. The coefficient of isotropic permeability of the left-half of the aquifer K_1 is three times that of the right-half of the aquifer K_2, that is, $K_1 = 3K_2$. Other pertinent data are given in the sketch. Determine the following: (i) pressure at point P and (ii) the coefficients of permeability, if the steady discharge per meter width of aquifer (normal to the plane of paper) is 1.0×10^{-5} m²/s.

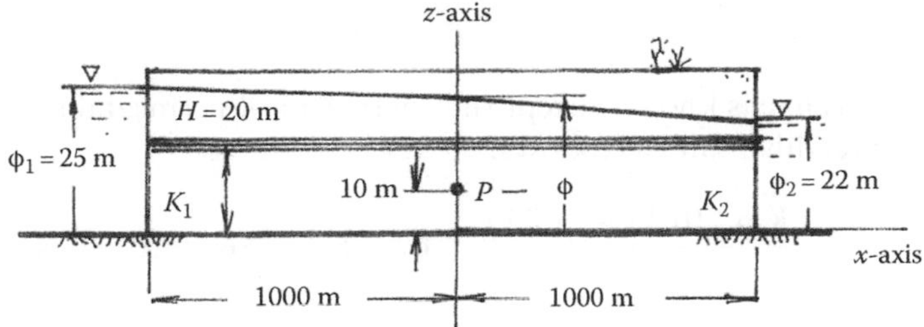

(i) *Pressure at Point P*: We assume that the piezometric head at the mid-section of the aquifer is ϕ. The discharge per meter width through the left-half of the aquifer is given as

$$Q_x(\text{left-half}) = \frac{K_1 H(\phi_1 - \phi)}{L} \tag{IP4.1.1}$$

Similarly, the discharge through the right-half of the aquifer is given by the following:

$$Q_x(\text{right-half}) = \frac{K_2 H(\phi - \phi_2)}{L} \tag{IP4.1.2}$$

Since the flow is in a steady state, with no sink or source within the aquifer, the two discharges given earlier must be equal. Thus, equating the right-hand sides of the preceding two equations yields the following:

$$K_1(\phi_1 - \phi) = K_2(\phi - \phi_2) \tag{IP4.1.3}$$

It is given that $K_1 = 3K_2$. Thus, substituting the value of K_1 in terms of K_2 in the preceding equation yields, after some rearranging and cancelling the terms, the following expression for the unknown ϕ:

$$\phi = \frac{(3\phi_1 + \phi_2)}{4} = \frac{(3 \times 25 + 22)}{4} = 24.25 \text{ m.} \tag{IP4.1.4}$$

From Equation IP4.1.4, the pressure head at P can be found as follows:

$$\frac{p}{\gamma} = \phi - z = 24.25 - 10 = 14.25 \text{ m.} \tag{IP4.1.5}$$

(ii) *Coefficients of permeability*: In the formula for discharge

$$Q_x = \frac{K_1 H(\phi_1 - \phi)}{L} \tag{IP4.1.6}$$

every variable is known except the coefficient of permeability K_1. This can be readily found as follows:

$$1.0 \times 10^{-5} = \frac{K_1 \times 20(25 - 24.25)}{1000}; \quad \text{or} \quad K_1 = 6.667 \times 10^{-3} \text{ m/s.} \tag{IP4.1.7}$$

Finally,

$$K_2 = \frac{K_1}{3}; \quad \text{or} \quad K_2 = \frac{(6.667 \times 10^{-3})}{3.0} = 2.222 \times 10^{-3} \text{ m/s.} \tag{IP4.1.8}$$

Remarks: The piezometric head, or the pressure head, at point P does not depend on the actual values of H, or of discharge Q_x, or of the coefficients of permeability, so long as the ratio K_1/K_2 remains the same.

4.2 A two-dimensional flow field (in x, z-plane) through a shallow horizontal, unconfined aquifer is shown in the following sketch. Using the Dupuit–Forchheimer assumptions, find the following: (i) the pressure at point P and (ii) the discharge per meter width of aquifer if the coefficient of permeability is $K = 0.0001$ m/s. Assume an isotropic, homogeneous aquifer with constant head boundaries at the left and right extremities.

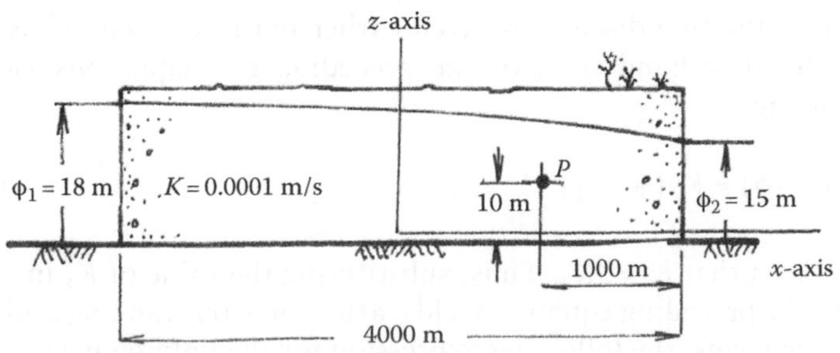

(i) *Pressure at P*: Based on the Dupuit–Forchheimer formulation, one obtains the following expression for the variation of the discharge potential for this problem:

$$\Phi = -(\Phi_1 - \Phi_2)\frac{x}{L} + \frac{1}{2}(\Phi_1 + \Phi_2); \quad \Phi = \frac{1}{2}K\phi^2 \tag{IP4.2.1}$$

The preceding equation can also be written in terms of the piezometric head ϕ, as follows:

$$\phi^2 = -(\phi_1^2 - \phi_2^2)\frac{x}{L} + \frac{1}{2}(\phi_1^2 + \phi_2^2) \tag{IP4.2.2}$$

Let at point P, $x = x_p$ and $\phi = \phi_p$, so that the preceding equation yields the following:

$$\phi_p^2 = -(\phi_1^2 - \phi_2^2)\frac{x_p}{L} + \frac{1}{2}(\phi_1^2 + \phi_2^2) \tag{IP4.2.3}$$

From the sketch, it can be readily found that $\phi_1^2 = 324\,\text{m}^2$, $\phi_2^2 = 225\,\text{m}^2$, and $x_p/L = 1/4$. Substituting these values in the previous equation yields the following:

$$\phi_p^2 = 249.75\ \text{m}^2; \quad \text{or} \quad \phi_p = 15.8\ \text{m}.$$

Thus, the pressure head at point P is given by the following:

$$\frac{p}{\gamma} = \phi_p - z_p = 15.8 - 10 = 5.8\ \text{m}. \tag{IP4.2.4}$$

(ii) *Discharge Q_x*: The discharge per foot width of the aquifer normal to the plane of paper is obtained as

$$\begin{aligned} Q_x &= -\frac{d\Phi}{dx} = \frac{(\Phi_1 - \Phi_2)}{L} = \frac{1}{2}\frac{K(\phi_1^2 - \phi_2^2)}{L} \\ &= \frac{0.0001(324 - 225)}{2 \times 4000} = 1.238 \times 10^{-6}\ \text{m}^2/\text{s}. \end{aligned} \tag{IP4.2.5}$$

Remarks: As before, the pressure, or the piezometric head, at point P does not depend on the actual value of K. It is independent of K. The question to ponder is: What happens to pressure when K becomes zero?

4.3 A well is centrally located in a circular island surrounded by constant-head boundary as shown in the following sketch. Assuming an isotropic confined aquifer of depth $H = 20$ m and the flow in a steady-state condition, find the coefficient of permeability K if the steady well discharge is $Q=0.01$ m³/s. Other pertinent data are given in the following figure.

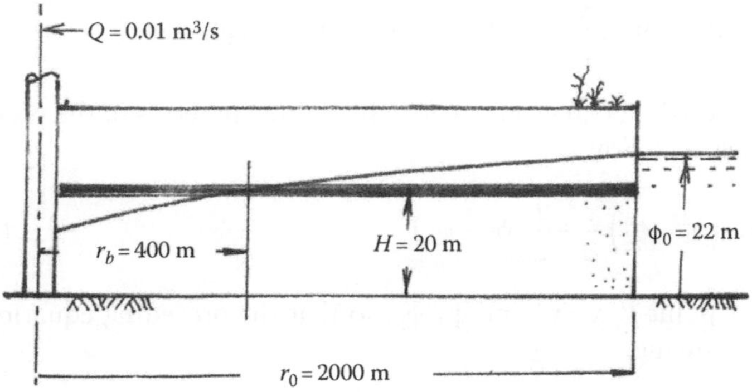

The formula for the interzonal boundary from the text is

$$\frac{r_b}{r_0} = \exp\left[-\left(\frac{2\pi H^2}{Q}K\right)\left(\frac{\phi_0}{H}-1\right)\right] \tag{IP4.3.1}$$

where

$H=20$ m

$\phi_0 = 22$ m

$Q=0.01$ m³/s

$$\frac{r_b}{r_0} = \frac{400 \text{ m}}{2000 \text{ m}} = \frac{1}{5}$$

Substituting the preceding values in Equation IP4.3.1 yields the following expression for K:

$$K = \frac{\ln\left(\frac{1}{5}\right)}{-\left(\frac{2\pi \times 400}{0.01}\right)\left(\frac{22}{20}-1\right)} = 6.4 \times 10^{-5} \text{ m/s.}$$

4.6 EXERCISES

4.1 The piezometric heads in three observation wells are shown in the sketch. Assuming a homogeneous, isotropic confined aquifer of depth 12 m, find the discharge potential $\Phi(x, y)$ as a function of x and y if there are no sources or sinks in the vicinity of wells. You may assume that the aquifer is shallow and horizontal in x, y-plane. Find Q_x, Q_y, and the maximum discharge Q_{max} (discharge per width through the entire depth of aquifer). The coefficient of permeability is given as $K = 0.0005$ m/s. (*Hint*: The discharge potential may be assumed as a linear function of x and y and, consequently, the flow may be assumed rectilinear. A reference to Figure 4.1 in the text may be helpful.)

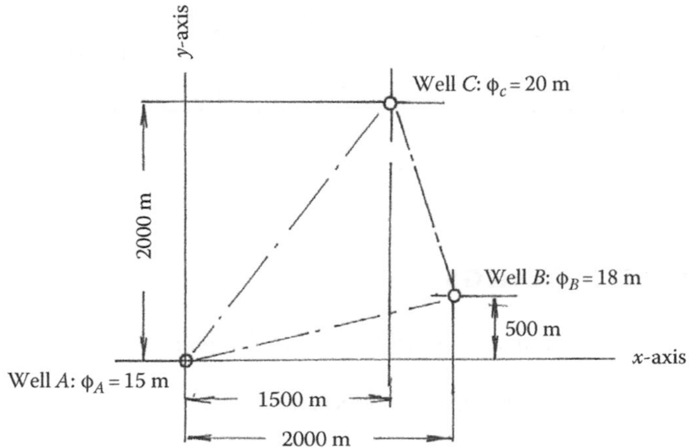

4.2 Treating as a shallow, horizontal unconfined aquifer, solve Excercise 4.1, using Dupuit–Forchheimer's assumption. Disregard the aquifer depth, with the exception of other pertinent data, in the preceding exercise.

4.3 The following figure describes one-dimensional flow through a partly confined aquifer. Find the location of the interzonal boundary. The aquifer material is homogeneous, isotropic with coefficient of permeability $K = 0.0005$ m/s. Other data are shown on the figure. What is the discharge Q_x?

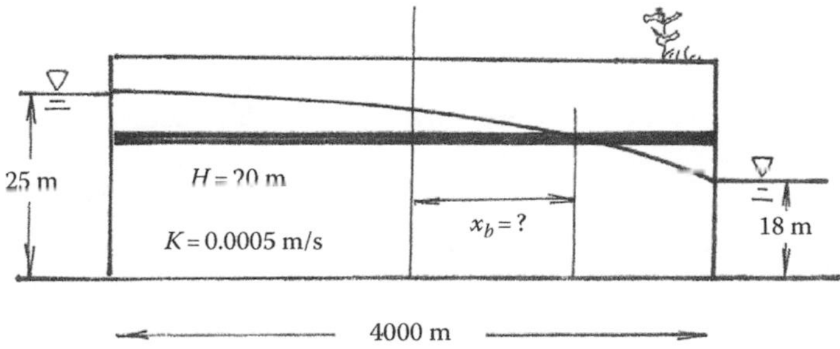

4.4 The following figure describes the radial flow toward a well in a partly confined aquifer. Find well discharge Q and corresponding r_b. Pertinent data are given on the figure. Assume isotropic homogeneous aquifer material with $K = 0.0005$ m/s.

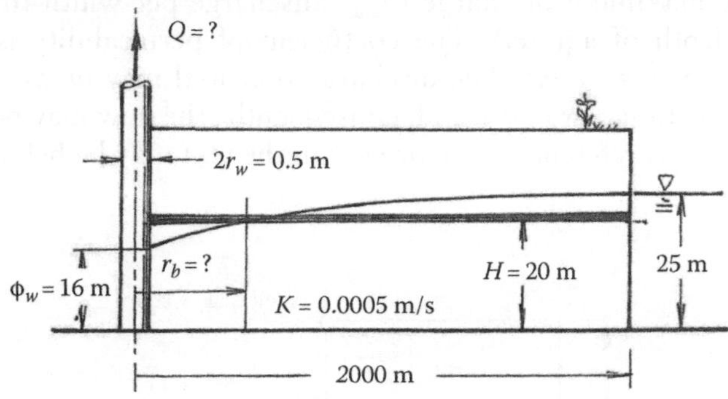

SUGGESTED READINGS

Hermance, J. F. 1999. *A Mathematical Primer on Groundwater Flow*, Prentice Hall, Upper Saddle River, NJ.

Strack, O. D. L. 1989. *Groundwater Mechanics*, Prentice Hall, Inc., Englewood Cliffs, NJ.

Todd, D. K. 1959. *Ground Water Hydrology*, John Wiley & Sons, Inc., New York.

Chapter 5

Laplace equation, superposition of harmonic functions, and method of images

The solutions of Laplace equation are called harmonic functions. These harmonic functions enjoy certain properties that are quite useful in obtaining a variety of solutions to practical problems. Here, we are interested in solutions to a number of engineering problems related to the movement of groundwater. In this regard, the so-called *method of images* is of immense usefulness in solving the boundary-value problems governed by the Laplace equation—especially when the boundary conditions are those that are often encountered in practice, but are not readily amenable to analysis.

5.1 SOME IMPORTANT PROPERTIES OF HARMONIC FUNCTIONS

Before we start describing the method of images, it is imperative that we discuss informally the Laplace equation and its solutions in general terms. Those who are interested in a greater precision and mathematical rigor may refer, for instance, to Kellogg (1953) and Sneddon (1957). The following informal assertions about the subject matter from our perspective are quite adequate for the time being.

Assertion 5.1: If Φ_1 and Φ_2 are two solutions of the Laplace equation, then their linear combination $\Phi = c_1\Phi_1 + c_2\Phi_2$ is also the solution of the Laplace equation, where c_1 and c_2 are arbitrary constants. This assertion is true because the Laplace equation,

$$\nabla^2\Phi(x, y) \equiv \frac{\partial^2\Phi}{\partial x^2} + \frac{\partial^2\Phi}{\partial y^2} = 0,$$

is a linear homogeneous differential equation. The truth of this assertion can be easily demonstrated by substitution of $\Phi = c_1\Phi_1 + c_2\Phi_2$ in the Laplace equation and making use of the fact that Φ_1 and Φ_2 are harmonic functions. For instance,

$$\nabla^2\left(c_1\Phi_1 + c_2\Phi_2\right) = c_1\nabla^2\Phi_1 + c_2\nabla^2\Phi_2 = 0 + 0 = 0$$

The following are the special cases of linear combinations: $\Phi = \Phi_1 + \Phi_2$, $\Phi = c_1\Phi_1$, and $\Phi = c_2\Phi_2$. Incidentally, this assertion is also true for a linear combination of a number of harmonic functions, greater than two, such as

$$\Phi = c_1\Phi_1 + c_2\Phi_2 + \cdots + c_n\Phi_n.$$

Assertion 5.2: The maximum or the minimum value of the harmonic function $\Phi(x, y)$ always lies on the boundary of the region.

Assertion 5.3: In a region (bounded by one continuous boundary, or multiple continuous boundaries), the harmonic function $\Phi(x, y)$ is uniquely defined for the following types of boundary conditions:

- *Type 1 (Dirichlet's problem)*: Everywhere on the boundary the potential $\Phi(x, y)$ is specified.
- *Type 2 (Neumann's problem)*: Everywhere on the boundary, the derivative $\partial\Phi/\partial n$ along the outward normal to the boundary is specified.
- *Type 3 (Mixed boundary-value problem)*: In this case, the Type 1 boundary condition is specified on the part of the boundary and the Type 2 boundary condition is specified on the remaining part of the boundary.

Type 3 is the most general and it contains the remaining two types as particular cases. Thus, the existence of a unique potential inside the region for a Type 3 boundary condition also implies the uniqueness of $\Phi(x, y)$ in the remaining two cases.

Assertion 5.4: The value of a harmonic function at a given point, (x, y), is always equal to the mean value of the function on the circumference of a circle with its center at the point (x, y). The radius of the circle is however arbitrary.

Finally, whereas *Assertion 5.1* reveals the possibility of multiple harmonic functions in a given region, *Assertion 5.3* limits the multiplicity to a unique harmonic function under three types of boundary conditions.

5.2 METHOD OF IMAGES

The method of images is a skillful way of employing the solutions of the Laplace equation, valid for an infinite domain, to the practical problems defined generally over a finite (or semi-finite) domain. The method has been used previously in the field of electrostatics and heat conduction in physics. It depends on the superposition of a finite number of harmonic functions, such that a part of the superposed solution of the Laplace equation satisfies the specified boundary conditions of a given practical problem. Thus, based on *Assertion 5.3* stated earlier, the part of the superposed solution of the Laplace equation also uniquely defines the actual solution of the given problem, if the superposed part happens to meet the actual boundary conditions imposed by a given physical problem. The details of the method are better understood by an example problem, as shown in the following sections.

5.2.1 Well at a finite distance from an infinitely long stream

Our goal in discussing this problem is to demonstrate how the mathematical solution of potential function for the flow field around a well operating in the vicinity of a long perennial river can be obtained by simple superposition of elementary harmonic functions. For this purpose, *we take the potential function, Φ, of flow field around a fully penetrating well in a confined aquifer of infinite extent as the elementary harmonic function.* We choose two wells located $2d$-distance apart as shown in Figure 5.1. In this figure, one of the wells is the regular *discharge well* with discharge Q, while the other is a *recharge well* with discharge $-Q$. The regular well pumps water out of the aquifer, while the recharge well pumps water into the aquifer. Throughout this discussion, we shall assume a horizontal, homogeneous, isotropic confined aquifer of extensive areal extent with a coefficient of permeability K.

The discharge potential for a fully penetrating well in a horizontal confined aquifer of extensive areal extent is given by the following equation:

$$\Phi_1 = \frac{Q}{2\pi}\ln r_1 + C_1 \tag{5.1}$$

where

Q is the steady (not changing with time) well discharge

The variable r_1 is the radial distance from the well to a movable point P(x, y) where Φ_1 is evaluated

C_1 is a constant

It is apparent from Equation 5.1 that C_1 represents the potential at a distance $r_1 = 1$. Since potential is always measured with respect to an arbitrary datum, it becomes obvious that C_1 is an arbitrary constant. Let

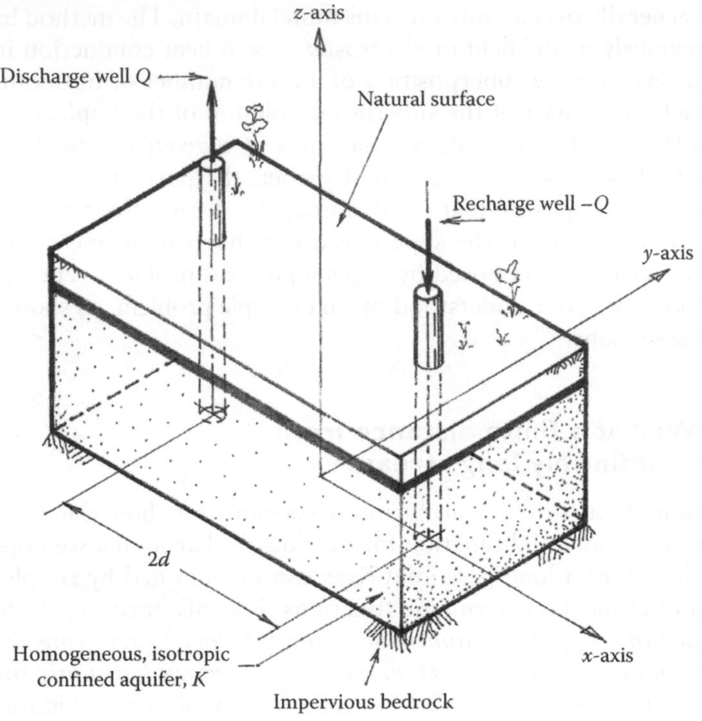

Figure 5.1 Definition sketch.

$$\Phi_2 = \frac{-Q}{2\pi}\ln r_2 + C_2 \tag{5.2}$$

be the discharge potential of a *recharge* well (discharge equal to $-Q$), where r_2 is the radial distance from the recharge well to the point P(x, y), and C_2 is another arbitrary constant. If both the *discharge well* and the *recharge well* are simultaneously operating in the same horizontal, confined aquifer of extensive areal extent, the combined discharge potential Φ, according to *Assertion 5.1*, is given by the following linear combination:

$$\Phi = \Phi_1 + \Phi_2 = \frac{Q}{2\pi}\left(\ln r_1 - \ln r_2\right) + C_1 + C_2 \tag{5.3a}$$

$$\Phi = \frac{Q}{2\pi}\ln\frac{r_1}{r_2} + C \tag{5.3b}$$

where $C = C_1 + C_2$ is an arbitrary constant. The resulting flow field is shown in Figure 5.2. It is quickly seen from this figure that $r_1 = r_2$ represents the

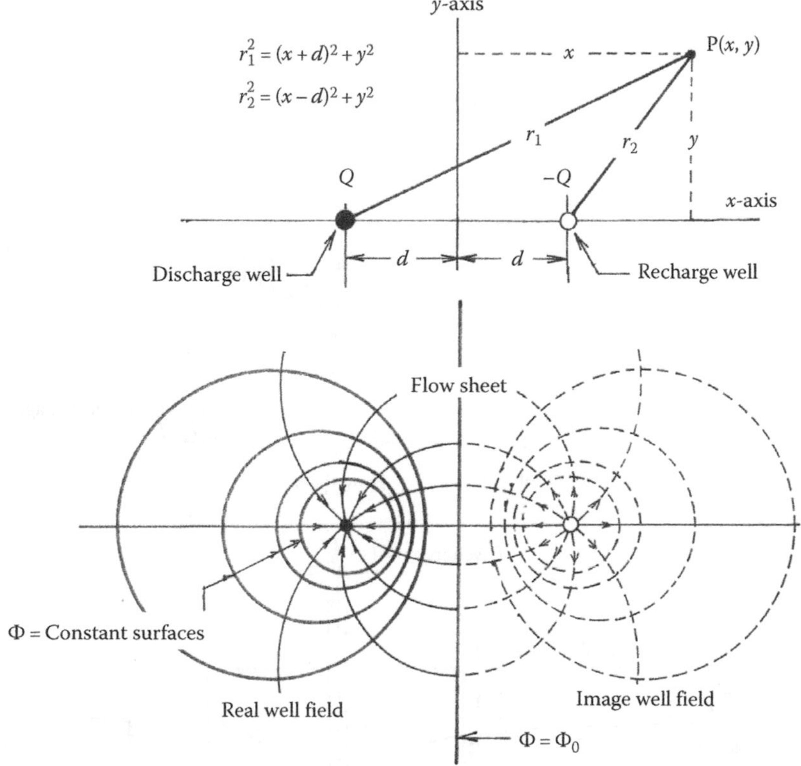

Figure 5.2 Superposition of two elementary harmonic functions.

locus of the right-bisector (i.e., a line that bisects a given line segment at right-angle) of the line joining the two wells. Hence, along this locus (y-axis in Figure 5.2), the discharge potential remains constant, that is,

$$\Phi = \Phi_0 = C \tag{5.4}$$

Now, if the flow field on the right-hand side of the y-axis, including the recharge well, is removed and replaced by some equivalent constant head boundary along the y axis—such as a fully penetrating vertical bank of a long perennial river, the effect of this change, mathematically speaking, will not even be felt by the flow field on the left-hand side of the y-axis. In other words, according to *Assertion 5.3*, the mathematical solution on the left side of the y-axis can also represent a unique harmonic solution of the flow field due to a *single* discharge well, operating in *isolation* at a distance d from an infinitely long perennial river with constant head. This situation is shown in Figure 5.3. In this figure, the y-axis represents

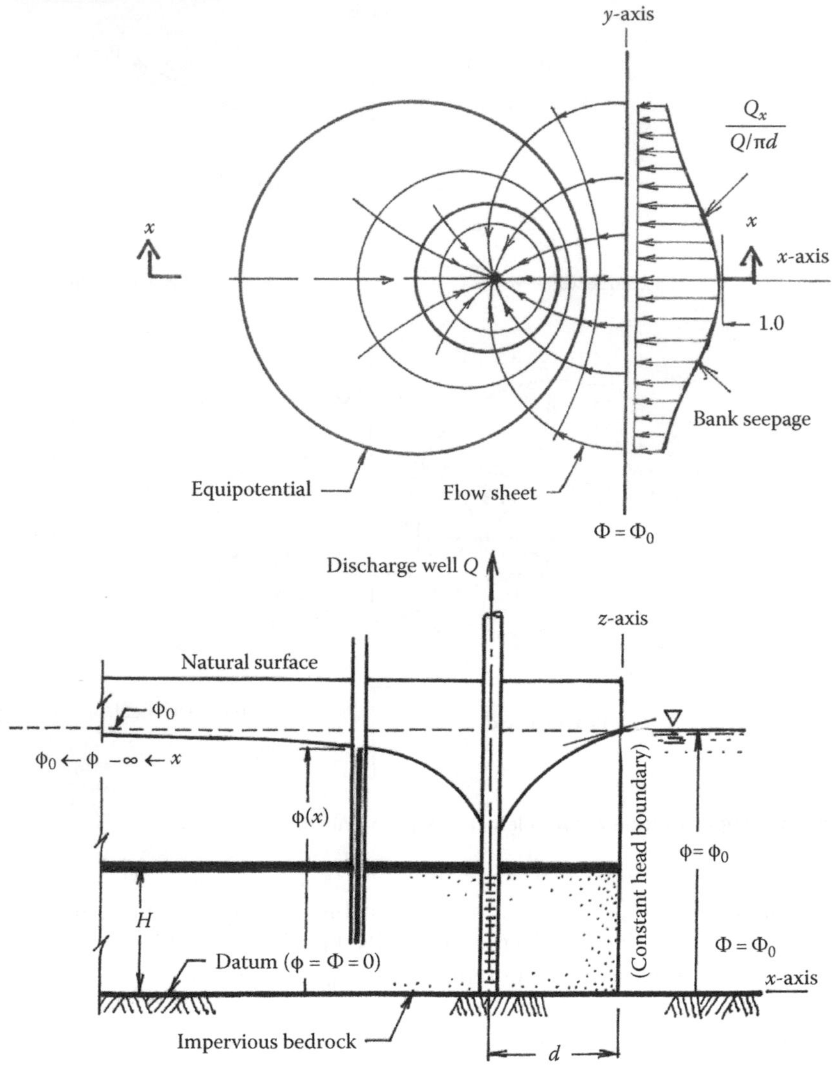

Figure 5.3 A well at a finite distance from an infinitely long perennial river.

the fully penetrating vertical bank of a perennial river with constant head boundary condition, that is, $\phi = \phi_0$ and consequently $\Phi = \Phi_0 (= KH\phi_0)$ as shown in the figure.

The recharge well does not exist in reality and is only used as a mathematical contraption to obtain a harmonic function that satisfies the actual boundary condition of a physical problem. In the literature, the recharge well is also called the *image well*, because it represents the image of the actual well if the constant-head boundary (e.g., y-axis) is regarded as a mirror. Of course, the mirror in this case is somewhat *funny*, for it changes the discharge Q of the real well into discharge $-Q$ of the image well.

What we have hitherto described is the essence of the *method of images*. The method is, however, more general and can be applied to obtain solutions to other problems of practical importance. Further applications are discussed later in this chapter.

In order to highlight the essential features of the flow field, Figure 5.4 has been prepared. With regard to this figure, it is worth mentioning that both the discharge and the recharge wells fully penetrate the confined aquifer, and the resulting flow field is truly two-dimensional in the x, y-plane, with no variation along the z-axis. All surfaces depicting the equipotentials and the flow sheets are circular cylindrical surfaces, with axes of cylinders parallel to the z-axis. This is mathematically true for horizontal confined aquifers of extensive areal extent. Incidentally, the depth H, or the coefficient of permeability K, of the confined aquifer does not influence the circular cylindrical nature of these surfaces, so long as the wells are fully penetrating, depth H remains finite, and the areal extent of the aquifer remains infinite. Furthermore, equipotential surfaces and flow sheets constitute a set of orthogonal family, because the aquifer is isotropic.

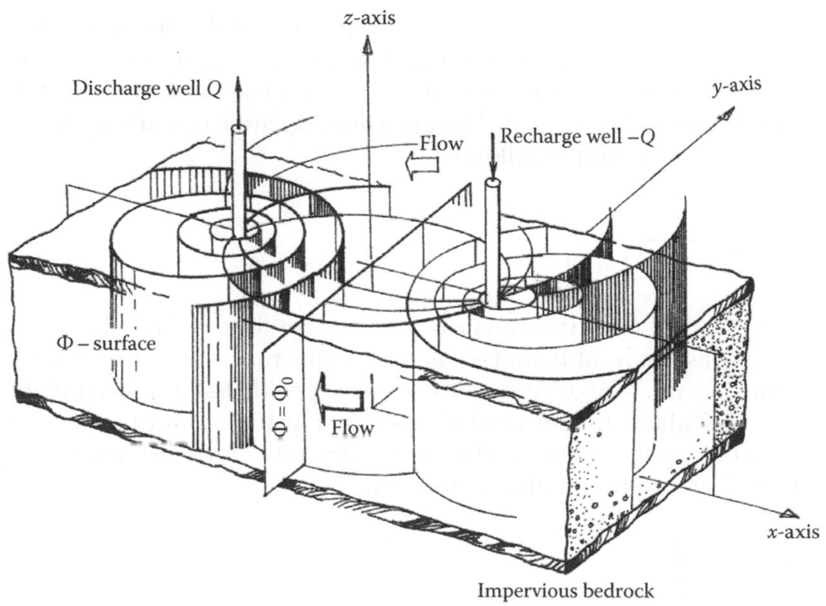

Figure 5.4 Perspective view of superposition of two elementary harmonic functions.

Combined discharge potential in rectangular Cartesian coordinate system: For further analysis of the flow field around a single well operating in the vicinity of a long perennial river, it is more convenient to work with the rectangular coordinate system. For this purpose, a reference to Figure 5.2 will be useful. The Equation 5.3a can be expressed in terms of x, y coordinates as shown in the following:

$$\Phi(x,y) = \frac{Q}{4\pi}\left[\ln\left\{(x+d)^2 + y^2\right\} - \ln\left\{(x-d)^2 + y^2\right\}\right] + \Phi_0 \tag{5.5}$$

The discharge Q_x can be easily obtained by differentiating the preceding equation as follows:

$$Q_x = -\frac{\partial \Phi}{\partial x} = -\frac{Q}{4\pi}\left[\frac{2(x+d)}{\left\{(x+d)^2 + y^2\right\}} - \frac{2(x-d)}{\left\{(x-d)^2 + y^2\right\}}\right] \tag{5.6}$$

The variation of discharge Q_x along the y-axis can be determined by substituting $x=0$ in the preceding equation to obtain

$$Q_x = -\frac{Q}{\pi d}\left[\frac{1}{1+(y/d)^2}\right] \tag{5.7a}$$

which shows that Q_x acts always in the negative x-direction, along the y-axis. In other words, there is seepage from the river into the confined aquifer. Furthermore, the amount of seepage per length of river is a function of three variables, Q, y, d. The preceding equation can also be written in a dimensionless form as follows:

$$\frac{Q_x}{Q/\pi d} = -\left[\frac{1}{1+(y/d)^2}\right] \tag{5.7b}$$

This shows that the dimensionless seepage per length into the aquifer from the constant-head boundary is only a function of the dimensionless y-coordinate, y/d. The dimensionless seepage profile has also been plotted in Figure 5.3 along a segment of the y-axis. Equation 5.7a can further be integrated along the y-axis to obtain the time rate of total seepage, Q_{total}, from the river into the aquifer, as follows:

$$Q_{total} = 2\int_0^\infty Q_x \, dy$$

where, Q_{total} is the total seepage volume per time, [L^3/T]

It is left as an exercise for the reader to prove that $Q_{total}=Q$, where Q is the steady well discharge [L^3/T]. This result, incidentally, also demonstrates that the solution obtained by the *method of images* for a single well operating in the vicinity of a long river *with constant head* does indeed represent the steady-state condition.

Geometry of equipotential curves: From Equation 5.5, it is possible to infer analytically the geometrical features of the equipotential curves. For this purpose, Equation 5.5 can be rearranged to obtain the following form:

$$\frac{\Phi(x,y)-\Phi_0}{Q/4\pi}=\ln\left[\frac{(x+d)^2+y^2}{(x-d)^2+y^2}\right] \tag{5.8}$$

where the term on the left-hand side, $\dfrac{\Phi(x,y)-\Phi_0}{Q/4\pi}$, represents the dimensionless potential relative to Φ_0. The constant Φ_0 represents the discharge potential at points determined by $r_1=r_2$. For a given value of well discharge, Q, the potential $\Phi(x, y)=\text{const}$ represents a curve in the x, y-plane. The nature of this curve can be obtained by setting the left-hand side of Equation 5.8 equal to a constant. Let this constant be denoted by

$$\theta\equiv\frac{\Phi(x,y)-\Phi_0}{Q/4\pi}$$

so that Equation 5.8 can be written in the following form:

$$e^{\theta}=\frac{(x+d)^2+y^2}{(x-d)^2+y^2} \tag{5.9a}$$

or

$$e^{\theta}\left[(x-d)^2+y^2\right]=\left[(x+d)^2+y^2\right] \tag{5.9b}$$

After some algebraic manipulations, the preceding equation can be cast in the following form.

$$-d^2=x^2-2xd\left[\frac{e^{\theta}+1}{e^{\theta}-1}\right]+y^2 \tag{5.10}$$

By adding $[(e^{\theta}+1)/(e^{\theta}-1)]^2d^2$ to both sides of Equation 5.10, we can obtain the following equation:

$$\left[\left(\frac{e^{\theta}+1}{e^{\theta}-1}\right)^{2}-1\right]d^{2}=\left[x-\left(\frac{e^{\theta}+1}{e^{\theta}-1}\right)d\right]^{2}+y^{2} \tag{5.11}$$

For a constant value of θ, Equation 5.11 represents a circle with center at

$$x_c=\left(\frac{e^{\theta}+1}{e^{\theta}-1}\right)d \tag{5.12a}$$

$$y_c=0 \tag{5.12b}$$

and of radius

$$R=+\sqrt{\left[\left(\frac{e^{\theta}+1}{e^{\theta}-1}\right)^{2}-1\right]d^{2}} \tag{5.12c}$$

The preceding equations can be expressed in a more convenient form utilizing the normalized coordinates, x/d, y/d and R/d with the result

$$\frac{x_c}{d}=\left(\frac{e^{\theta}+1}{e^{\theta}-1}\right) \tag{5.13a}$$

$$\frac{y_c}{d}=0 \tag{5.13b}$$

$$\frac{R}{d}=+\sqrt{\left[\left(\frac{e^{\theta}+1}{e^{\theta}-1}\right)^{2}-1\right]} \tag{5.13c}$$

The preceding results are graphically shown in Figure 5.5. In the lower sketch, the abscissa represents the normalized x-coordinate, while the ordinate represents both the normalized y-coordinate as well as the normalized radius of the equipotential circle. The real well is situated at $x/d=-1.0$ (in other words, the well is located at a distance d from the perennial river on the negative x-axis). The thick solid curve, identified by C-R in the figure, represents the relationship between the normalized x-coordinate of the center and the normalized radius of the equipotential circle. A selected number of four equipotential circles, with curve parameter

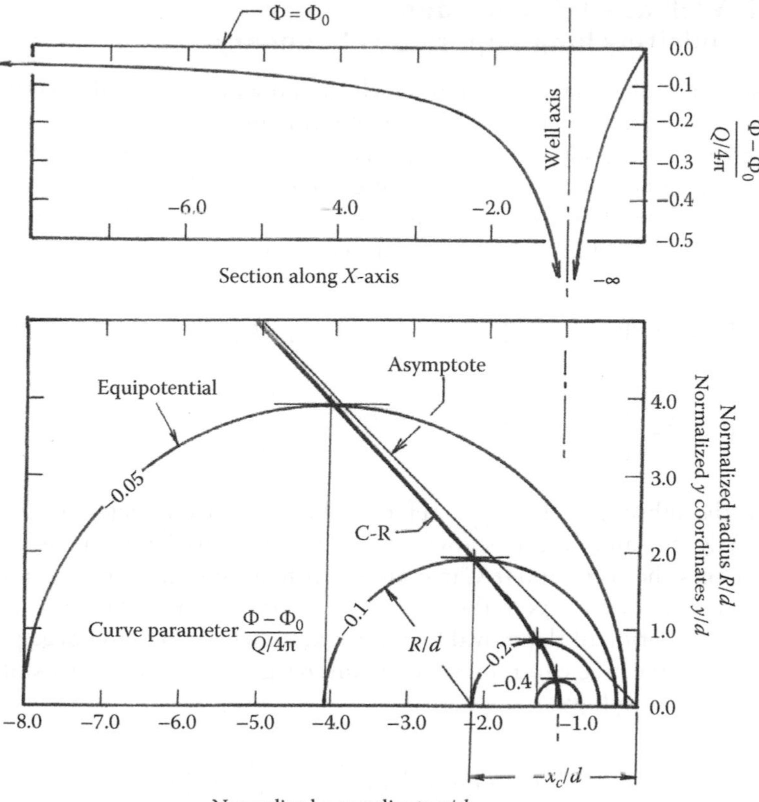

Figure 5.5 Plot of normalized potential and equipotential contours.

$$\theta \equiv \frac{\Phi(x,y)-\Phi_0}{Q/4\pi} = -0.05, -0.10, -0.20, \text{ and } -0.40$$

is shown in this figure. The numerical values representing the progression of curve parameter are so chosen that the center of each equipotential circle lies on the circumference of the next proceeding circle. The centers of proceeding equipotential circles tend toward the well axis with a decreasing value of curve parameter, θ, as shown in the sketch.

A cross section of the potential surface along the x-axis is shown in the upper sketch. It is evident from this sketch that $\Phi(x, y) = \Phi_0$ at a point where $x/d = 0$. The ordinate in this figure represents the dimensionless potential $\frac{\Phi(x,y)-\Phi_0}{Q/4\pi}$ relative to Φ_0, where Φ_0 represents the potential at the perennial river as well as at the far field (i.e., $x \to \pm\infty$ or $y \to \pm\infty$).

5.2.2 Well at a finite distance from an infinitely long impervious boundary

As a second application, we find the discharge potential for a well operating in the vicinity of a long, straight, fully penetrating impermeable boundary. For this purpose, we assume two fully penetrating wells, located at $2d$-distance apart with equal discharge Q, operating simultaneously in a confined aquifer of infinite areal extent. Following *Assertion 5.1*, the combined discharge potential can be obtained by superposition as follows:

$$\Phi(r_1, r_2) = \Phi_1 + \Phi_2 = \frac{Q}{2\pi}(\ln r_1 + \ln r_2) + C_1 + C_2 \tag{5.14a}$$

$$\Phi(r_1, r_2) = \frac{Q}{2\pi}\ln r_1 r_2 + C \tag{5.14b}$$

where the subscripts, 1 and 2, refer to the variables associated with the two wells, W_1 and W_2, as shown in Figure 5.6. This figure, incidentally, also defines the rectangular Cartesian coordinate system in relation to the location of two wells. As in the previous case, the radial distances r_1 and r_2 locate the position of the movable point P(x, y), where the discharge potential is evaluated. We can represent the discharge potential in terms of x, y coordinates as follows:

$$\Phi(x, y) = \frac{Q}{4\pi}\left[\ln\left\{(x+d)^2 + y^2\right\} + \ln\left\{(x-d)^2 + y^2\right\}\right] + C \tag{5.15}$$

Discharges Q_x and Q_y at any point P(x, y) can be obtained by differentiating the preceding equation with respect to x and y, respectively, that is,

$$Q_x(x, y) = -\frac{\partial\Phi}{\partial x} = -\frac{Q}{4\pi}\left[\frac{2(x+d)}{\left\{(x+d)^2 + y^2\right\}} + \frac{2(x-d)}{\left\{(x-d)^2 + y^2\right\}}\right] \tag{5.16a}$$

$$Q_y(x, y) = -\frac{\partial\Phi}{\partial y} = -\frac{Q}{4\pi}\left[\frac{2y}{\left\{(x+d)^2 + y^2\right\}} + \frac{2y}{\left\{(x-d)^2 + y^2\right\}}\right] \tag{5.16b}$$

It is evident that the potential function $\Phi(x, y)$ does not change when x is replaced by $-x$, or y is replaced by $-y$, in Equation 5.15. Thus, the constant *Φ–curves* are symmetrical about the y-axis, as well as about the x-axis.

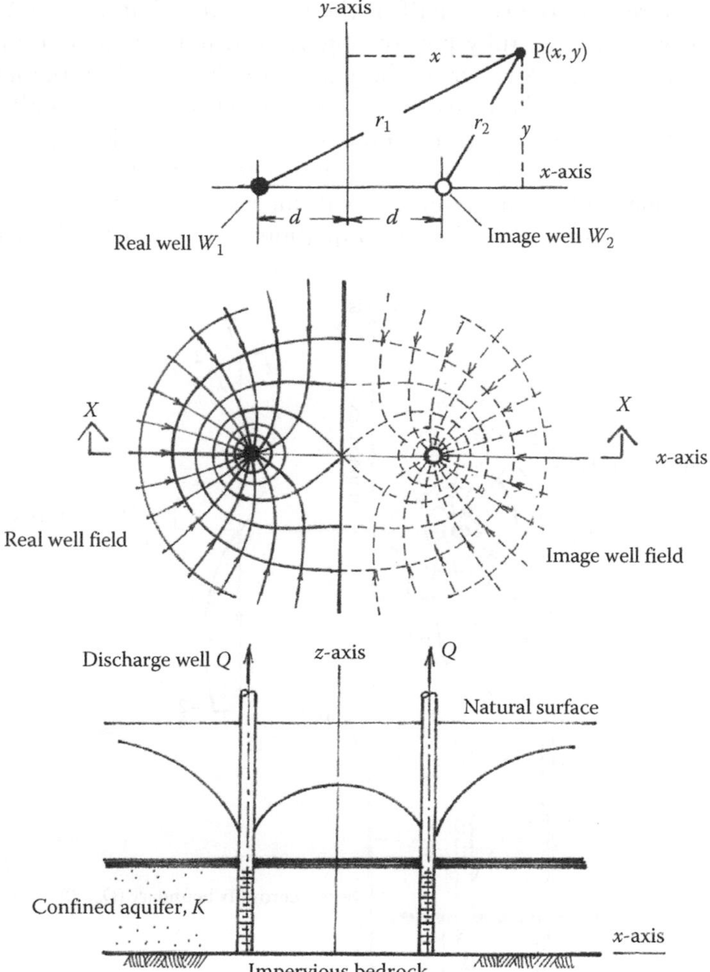

Figure 5.6 Superposition of two elementary functions.

In three dimensions, a similar symmetry is maintained when the term *curves* is replaced by *prismatic surfaces* and the y-axis by the y, z-plane and the x-axis by the x, z-plane.

By substituting $x=0$ in Equation 5.16a, it is clear that $Q_x=0$ along the y-axis. In other words, the combined harmonic function in this case is such that there is no flow across the right bisector of the line joining the two wells. As a conclusion, we state that in this case the specific discharge vector q at any point on the y-axis (or y, z-plane) acts parallel to the y-axis.

Now, if we remove the right half of the flow field including well W_2 and replace it with a long, fully penetrating vertical impervious boundary in order to ensure zero flow across the y-axis, the flow field on the left half ($x<0$) will not even feel the absence of the flow field on the right half ($x>0$). Thus, the flow field on the semi-infinite ($x<0$) plane can also represent the actual flow field due to a well operating in *isolation* at a distance d from an infinitely long, fully penetrating vertical impervious boundary. This situation is shown in Figure 5.7. From Equations 5.16a and 5.16b, it is also

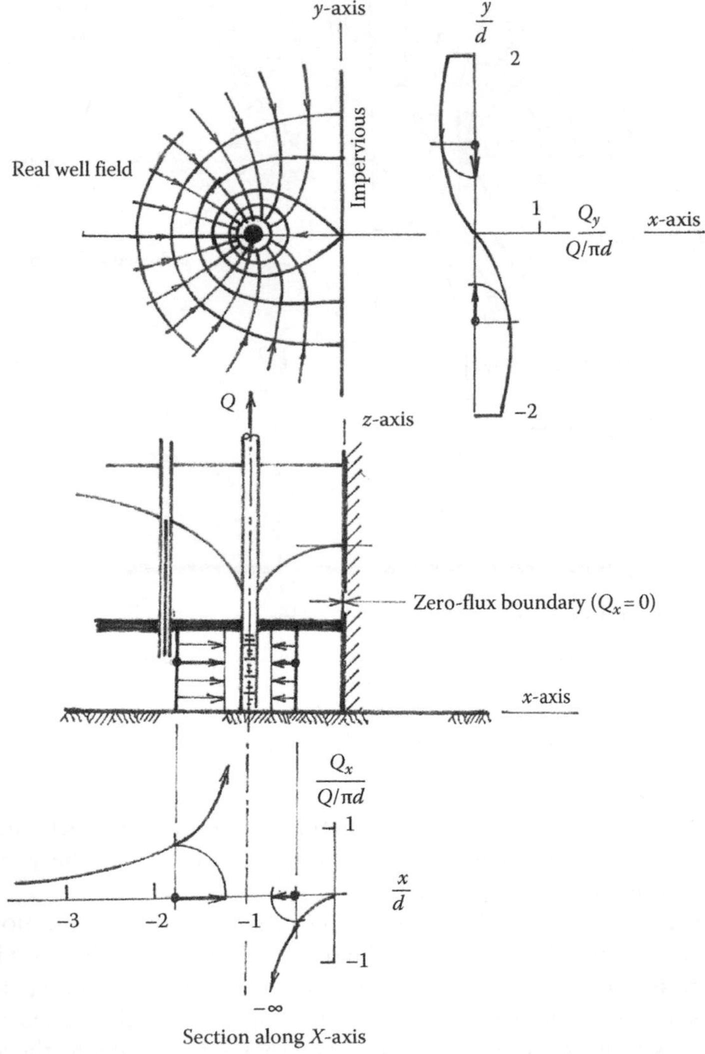

Figure 5.7 Well in the vicinity of impervious boundary.

possible to obtain the following two equations representing the dimensionless discharge along the x-axis and the y-axis, respectively:

$$\frac{Q_x}{Q/\pi d} = -\frac{x/d}{(x/d)^2 - 1} \tag{5.17a}$$

$$\frac{Q_y}{Q/\pi d} = -\frac{y/d}{(y/d)^2 + 1} \tag{5.17b}$$

These profiles of dimensionless discharges are also shown along the x-axis and the y-axis in Figure 5.7. It can be seen from this figure that $Q_x \rightarrow \pm\infty$ as x/d approaches the well axis. Also, along the y-axis the discharge Q_y always acts toward the origin that represents the stagnation point (a point where $q = 0$).

The well W_2 that is being removed is called the *image well* in the literature. It is not a real well and should be construed simply as a device to obtain the harmonic solution to the real boundary-value problem.

To obtain a mathematically acceptable solution to the flow field around a well operating at a distance d from an infinitely long, straight *impervious* boundary, we first find the image well by reflecting the actual well about a vertical reflecting plane surface located at the impervious boundary. Then, we assign the same discharge to the image well as that of the actual well. This is in contradistinction to the previous case of the well in the vicinity of a *constant-head* boundary, where the discharge of the image well was *negatively* equal to the discharge of the actual well.

In summary, the discharge of the real and image wells are *negatively equal to each other* in the case of a well operating in the vicinity of a straight constant-head boundary; otherwise, the discharge of the well and its image well are *equal to each other*, if the well operates in the vicinity of a straight impervious boundary.

5.2.3 Well operating in the vicinity of combined impervious and constant-head boundaries

The versatility of the method of images goes beyond the two cases discussed in the foregoing. It can be applied to cases such as shown in Figure 5.8. In all three cases shown in this figure, the well is operating in the vicinity of two *straight intersecting boundaries*. These boundaries could either represent the same type or different types of boundary conditions, for example, one representing the *constant-head* and the other representing the *impervious* boundary, as shown in Figure 5.8a. This figure is further discussed in some detail in the following text.

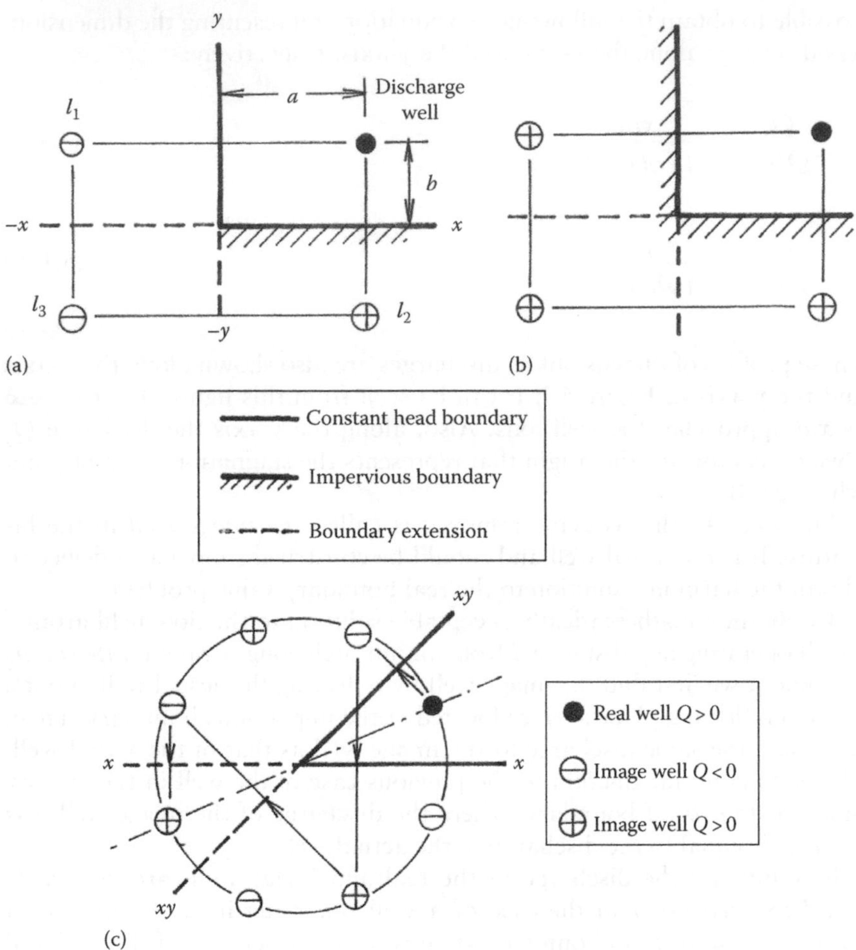

Figure 5.8 Aquifer bounded by two straight boundaries: (a) Constant-head and impervious boundaries intersecting at right-angle, (b) two impervious boundaries intersecting at right-angle, and (c) two constant-head boundaries intersecting at an angle of 45°.

Figure 5.8a represents a single discharge well operating in the vicinity of two different intersecting straight boundaries. The constant-head boundary is represented by a simple line, while the impervious boundary by a shaded (cross-hatched) straight line. The well, as shown by the solid circle in the figure, is located at a distance a from the constant-head boundary and at a distance b from the impervious boundary. The images of the operating well through the two boundaries are shown by open circles and are labeled as I_1 and I_2, respectively. It should be recalled that the discharge of the image well is negatively equal to the discharge of the well, when reflected through the constant-head boundary. Otherwise, the discharge of the image well remains the same as that of the well when reflected through the impervious boundary. The image identified by I_3 can be viewed in two different ways: It could be viewed either as the reflected image of the image well I_2 through the constant-head boundary (or its extension) or as the reflection of the image well I_1 through the impervious boundary (or its extension). It must however be kept in mind that each reflection through the constant-head boundary changes the sign of the discharge, while each reflection through the impervious boundary has no effect on the discharge of the reflected image. Thus, in order to obtain the flow field in the vicinity of the well bounded by two types of straight boundaries, one needs to combine the elementary solution of the well with the elementary solutions of three image wells, all operating in the same confined horizontal aquifer of infinite areal extent. The combined solution will of course meet the constant head boundary condition at the y-axis and the impervious boundary condition on the x-axis. To prove this statement, we proceed as follows. Let r, r_1, r_2, r_3 be the radial distances from the actual well and the three images I_1, I_2, I_3, respectively, to a movable point P(x, y). Thus, based on *Assertion 5.1*, the combined harmonic function can be expressed as follows:

$$\Phi = \frac{Q}{4\pi}\left[\ln\left(\frac{r^2 r_2^2}{r_1^2 r_3^2}\right)\right] + C \tag{5.18a}$$

or, equivalently, as

$$\Phi(x,y) = \frac{Q}{4\pi}\left[\ln r^2(x,y) + \ln r_2^2(x,y) - \ln r_1^2(x,y) - \ln r_3^2(x,y)\right] + C \tag{5.18b}$$

In Equation 5.18b, negative terms in the square bracket indicate that the discharges of the image wells, I_1, I_3 are negatively equal to Q, and C is an arbitrary constant. It can be readily seen that when the movable point P(x, y) lies on the y-axis, the sum of the terms inside the square bracket becomes zero, for $r = r_1$ and $r_2 = r_3$. In other words, combined harmonic function is such that the discharge potential Φ remains a constant, $\Phi = C(=\Phi_0)$, along the y-axis. This constant can appropriately be chosen to correspond with the constant head ϕ_0 in the perennial river. The important fact, however, is to note that Φ remains constant not only along the positive y-axis, but also along the negative y-axis, which represents the *extension of the boundary condition*, as shown in Figure 5.8 by a dotted line.

To demonstrate that the combined harmonic function also meets the impervious boundary condition on the x-axis, we need to calculate Q_y

along the x-axis. The discharge $Q_y(x, y)$ at any movable point $P(x, y)$ can be calculated by a simple partial differentiation of Equation 5.18b with respect to y, as follows:

$$Q_y = -\frac{\partial \Phi}{\partial y} = \frac{-Q}{4\pi}\left[\frac{1}{r^2}\frac{\partial r^2}{\partial y} + \frac{1}{r_2^2}\frac{\partial r_2^2}{\partial y} - \frac{1}{r_1^2}\frac{dr_1^2}{dy} - \frac{1}{r_3^2}\frac{\partial r_3^2}{\partial y}\right] \tag{5.19}$$

From Figure 5.9a, the following relationships can be seen:

$$r^2 = (x-a)^2 + (y-b)^2 \tag{5.20a}$$

$$r_1^2 = (x+a)^2 + (y-b)^2 \tag{5.20b}$$

$$r_2^2 = (x-a)^2 + (y+b)^2 \tag{5.20c}$$

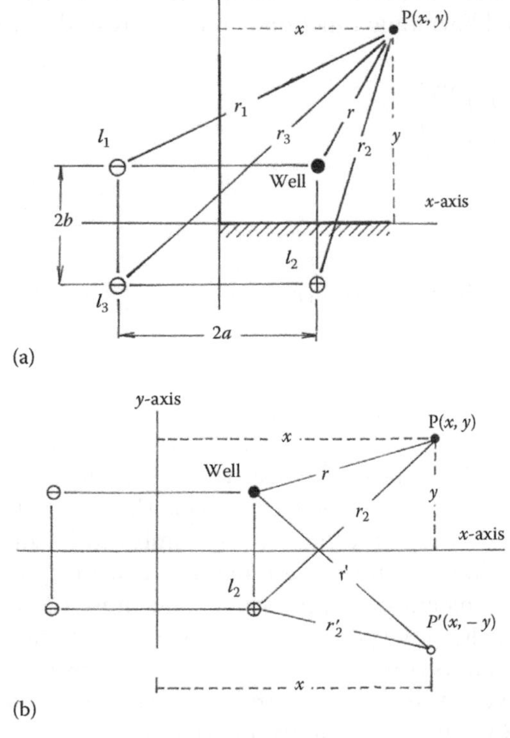

Figure 5.9 (a) Definition sketch and (b) invariance of product $r \cdot r_2$, under reflection through the x-axis of the variable point $P(x, y)$.

$$r_3^2 = (x+a)^2 + (y+b)^2 \tag{5.20d}$$

Differentiating the preceding relationships given in Equation 5.20 with respect to y, we obtain the following expressions:

$$\frac{\partial r^2}{\partial y} = 2(y-b) \tag{5.21a}$$

$$\frac{\partial r_1^2}{\partial y} = 2(y-b) \tag{5.21b}$$

$$\frac{\partial r_2^2}{\partial y} = 2(y+b) \tag{5.21c}$$

$$\frac{\partial r_3^2}{\partial y} = 2(y+b) \tag{5.21d}$$

Now combining Equation 5.19 with Equations 5.20 and 5.21 yields the following explicit expression for the discharge in terms of x, y coordinate:

$$Q_y(x,y) = \frac{Q}{4\pi}\left[\frac{2(y-b)}{(x-a)^2+(y-b)^2} + \frac{2(y+b)}{(x-a)^2+(y+b)^2} - \frac{2(y-b)}{(x+a)^2+(y-b)^2} - \frac{2(y+b)}{(x+a)^2+(y+b)^2}\right] \tag{5.22}$$

If we set $y=0$ in the preceding equation, we obtain the following result:

$$Q_y(x,0) = \frac{Q}{4\pi}[0] = 0 \tag{5.23}$$

which is valid for all points on the x-axis. Equation 5.23 shows that the combined harmonic function also satisfies the impervious boundary condition along the x-axis. Again, it is worth noting that the combined harmonic function satisfies the impervious boundary condition along the entire x-axis—including the extension shown by a dotted line in Figure 5.8.

We can, alternatively, demonstrate that the combined potential in this case satisfies the impervious boundary condition along the x-axis, by appealing to our geometric intuition. For this purpose, we refer to Figure 5.9b.

It is evident from this figure that the product $r \cdot r_2$ of the radial distances remains the same when the movable point P(x, y) is reflected through the x-axis to obtain its image $P'(x, -y)$. The same is of course true for the product $r_1 \cdot r_3$. Now, it follows from Equation 5.18a that the term inside the square bracket does not change as the movable point P(x, y) is replaced by its reflection through the x-axis. In other words, $\Phi(x, y) = \Phi(x, -y)$ or the potential function Φ is symmetric about the x-axis. Thus, the gradient vector of Φ, at points on the x-axis, is parallel to the axis. Since, in an isotropic medium, the specific discharge vector q is parallel to the gradient vector, it follows that, at all those points that lie on the x-axis, the specific discharge vector is parallel to the axis. Thus, the combined potential in this case leads to a flow field that creates its own *virtual impervious boundary* along the x-axis.

What we have accomplished so far is to find a harmonic function—defined throughout the x, y-plane—which also meets the constant-head boundary condition along the y-axis while maintaining the impervious boundary condition along the x-axis. Thus, this harmonic function meets all the requirements of the boundary-value problem in the first quadrant of the x, y-plane, and based on *Assertion 5.3* it is the unique solution to the potential field of a well operating in the vicinity of two intersecting straight boundaries.

The case shown in Figure 5.8b is either self-explanatory, or its explanation closely follows the previous discussion. A detailed discussion on this case is therefore omitted. The situation shown in Figure 5.8c, however, needs some explanation. This situation represents in practice the flow field of a well operating near the confluence of two perennial rivers. These rivers are idealized to represent two straight constant-head boundaries intersecting at an angle of 45°. In order to obtain a harmonic function that meets the constant-head boundary conditions at x-x line and xy-xy line, we need to combine the elementary solution of the well with other elementary solutions of seven image wells, as shown in the figure. The locations of image wells are obtained by reflecting alternatively the well through the two straight boundary lines. One sequence of such reflections is indicated by arrows in the figure. The other sequence of reflections can be similarly obtained by first reflecting the well through the x-x line. The other sequence of reflections is not shown for clarity.

5.2.4 Well between two parallel impervious boundaries

As a closing remark, it may be emphasized that a number of images may be finite or, in some cases, infinite depending upon the angle between the reflecting boundaries. As an example of the latter, consider the infinite sequence of images in the case of a well operating between two parallel

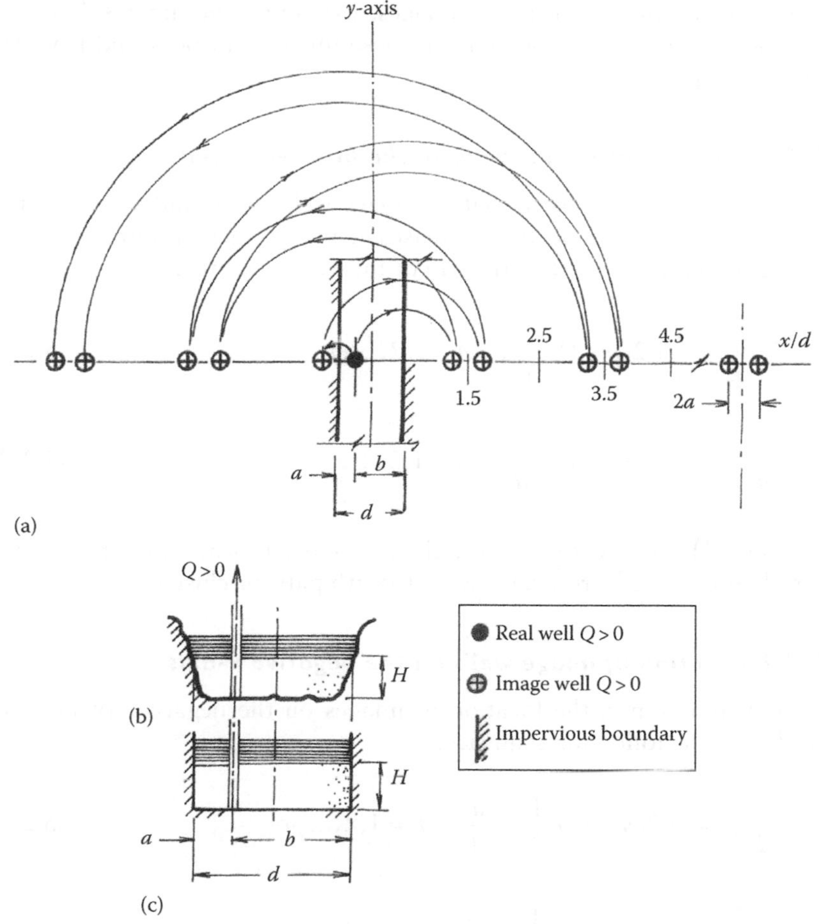

Figure 5.10 **(a) Plan view of well and its infinite sequence of images, (b) typical buried alluvial deposit in a straight river valley, and (c) an idealized uniformly confined aquifer.**

straight boundaries, as shown in Figure 5.10. This figure illustrates an idealized presentation of a buried alluvial deposit in a more-or-less straight river valley. In order to make it amenable to the method of images, the typical cross section of the buried deposit (Figure 5.10b) is replaced by a uniformly confined aquifer of width d and depth H. In the plan view (Figure 5.10a), the aquifer is represented by an infinite strip (extending along the y-axis) of a constant width d bound by two parallel impervious boundaries. The problem is to find the flow field in the vicinity of a single well located at distances a, b from the left and right impervious boundaries, respectively. In order to solve this problem using the method of images, one needs to consider an infinite sequence of images, extending on either side of the impervious boundaries. Locations of these images can be found from the following sequence.

5.2.4.1 Location of image wells on the positive x-axis

From Figure 5.10a, it is seen that the image wells are found in a cluster of two wells whose mean location increases in an arithmetic progression from the origin. Let n denote the nth term of this sequence, then

$$\left(\frac{x_L}{d}\right)_n = \frac{3}{2} + 2(n-1) - \frac{a}{d}; \quad n = 1,2,\ldots,\infty \tag{5.24a}$$

$$\left(\frac{x_R}{d}\right)_n = \frac{3}{2} + 2(n-1) + \frac{a}{d}; \quad n = 1,2,\ldots,\infty \tag{5.24b}$$

where $(x_L/d)_n$ and $(x_R/d)_n$ denote the normalized location of the left and the right image wells, respectively, in the nth pair (or cluster).

5.2.4.2 Location of image wells on the negative x-axis

In a similar manner, the location of images on the negative axis can be found from the following sequence:

$$\left(\frac{x_L}{d}\right)_n = -\left(2(n-1) + \frac{1}{2}\right) - \frac{a}{d}; \quad n = 1,2,\ldots,\infty \tag{5.25a}$$

$$\left(\frac{x_R}{d}\right)_n = -\left(2(n-1) + \frac{1}{2}\right) + \frac{a}{d}; \quad n = 1,2,\ldots,\infty \tag{5.25b}$$

In the preceding sequence, the term $n=1$ should be regarded with an exception, where $(x_R/d)_1$ in fact represents the actual well and not the image.

In practice, it is impossible to handle an infinite array of images. It, therefore, becomes necessary to replace the infinite sequence of image wells by a finite sequence of images close to the boundaries. If the solution converges rapidly, this truncation of infinite sequence works well; otherwise, the more sophisticated mathematical analysis based on the use of complex variables, or the numerical analysis utilizing the finite difference or the finite element method becomes necessary. The numerical methods are discussed later in this text; however, the application of complex variables, or theory of functions, is not within the prevue of this book. The interested reader may consult, for example, Rothe et al. (1961), Bear (1988), or Strack (1989).

5.3 METHOD OF IMAGES FOR CIRCULAR BOUNDARY

So far we have used the method of images with regard to *straight* constant-head or impervious boundaries. The genesis of this thought lies in the fact that the right-bisector of the line segment joining the well (with discharge Q) and its image well (with discharge $-Q$) represents the *straight* constant-head boundary—along which the piezometric head and consequently the discharge potential Φ remain constant. However, a review of Figure 5.2 will quickly reveal the fact that all equipotential curves, including the y-axis, along which Φ remains constant are circular in nature. The y-axis is indeed the circular arc of infinite radius. This naturally leads to the possibility that the method of images could also be extended to include flow regions with circular constant-head boundaries, such as a circular island surrounded by a large body of constant-head water. This situation is illustrated in Figure 5.11. Our objective here is to determine the flow field in the vicinity of a well that is eccentrically located with respect to the center of the circular island.

The top sketch in Figure 5.11 shows the circular island with radius R and the location of the pumping well. The middle sketch shows the cross section along the x-axis of the island and its surroundings. The bottom sketch shows the relevant geometrical parameters. We assume that the radius R and off-set δ of the well off the center of the island are part of the given data. Also, on the cylindrical surface at a distance R from the center, the (constant) piezometric head is known, that is, $\phi=\phi_0$, or $\Phi=\Phi_0(=KH\phi_0)$ on $x'^2+y'^2=R^2$. The x',y' coordinate system is located at the center of the island as shown in the lower sketch. To find the flow field in the vicinity of the eccentric well, we need the location, δ^*, of the image well, which can be obtained as shown subsequently. From the lower sketch of Figure 5.11, we find the following geometrical relations:

$$\delta = -x_c - d \tag{5.26a}$$

$$\delta^* = -x_c + d \tag{5.26b}$$

Multiplication of preceding two equations yields the following result:

$$\delta\delta^* = x_c^2 - d^2 \tag{5.26c}$$

Substituting the value of x_c from Equation 5.12a into the preceding equation yields

$$\delta\delta^* = \left(\frac{e^{\theta}+1}{e^{\theta}-1}\right)^2 d^2 - d^2 \tag{5.26d}$$

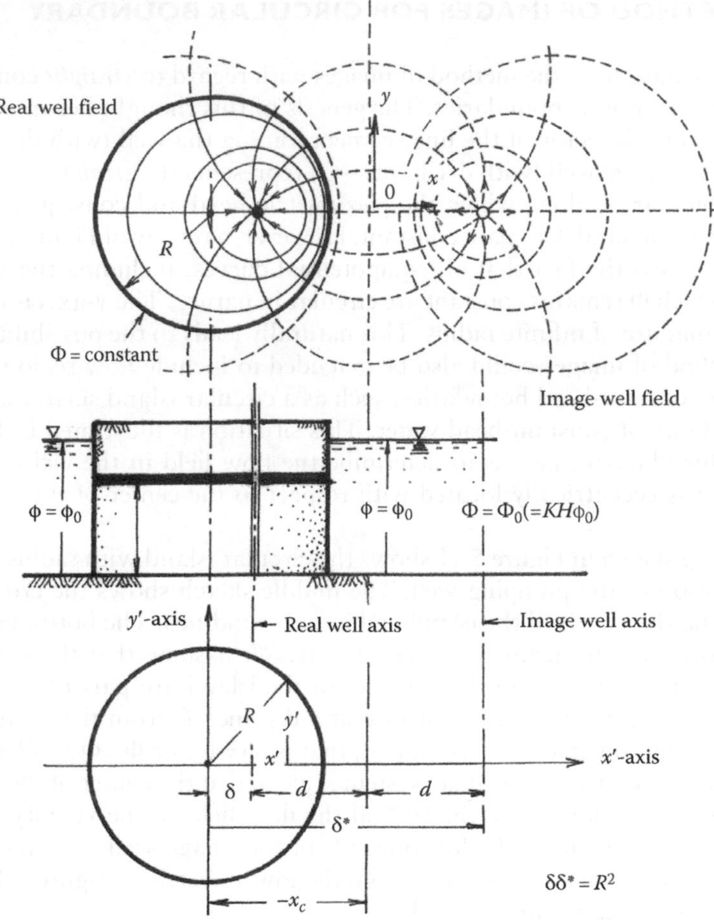

Figure 5.11 Eccentric well in a circular island.

By invoking Equation 5.12c, the preceding equation can be represented as follows:

$$\delta\delta^* = R^2 \tag{5.27}$$

from which the unknown

$$\delta^* = \frac{R^2}{\delta} \tag{5.28}$$

can be obtained in terms of the known data, R and δ.

5.4 ILLUSTRATIVE PROBLEMS

5.1 A well is located at 500 m from a long, straight constant-head boundary *aa*. The aquifer is uniformly isotropic with coefficient of permeability $K = 0.00014$ m/s. The steady-state piezometric levels at observation wells, A and B, are 20 m and 19 m, respectively, as shown in the following sketch. If the aquifer remains confined of uniform depth $H = 8$ m, find the following:

(a) The piezometric level in observation well located at P(1000, 500)
(b) The steady (not changing with time) discharge Q

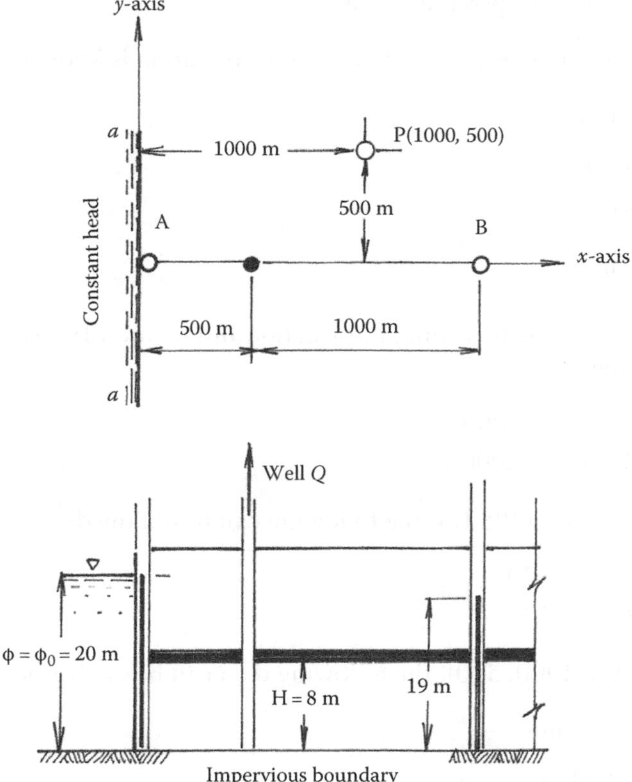

Part (a): This is a problem where the well is situated near a long constant-head boundary. The potential Φ is in general found from the following equation:

$$\Phi = \frac{Q}{2\pi} \ln\left[\frac{r_1}{r_2}\right] + \Phi_0 \tag{IP5.1.1}$$

Since the aquifer remains confined, the preceding equation can be replaced by the following equation:

$$\phi = \frac{Q}{2\pi KH} \ln\left[\frac{r_1}{r_2}\right] + \phi_0 \quad \text{(IP5.1.2)}$$

where

ϕ represents the piezometric level at the movable point $P(x, y)$

r_1 and r_2 represent the radial distances from the well and its image to the point $P(x, y)$, respectively

ϕ_0 represents the piezometric level in the observation well A (or the level in the perennial river)

At observation well B, the following information is known:

$r_1 = 1000$ m

$r_2 = 2000$ m

$\phi_B = 19$ m

$\phi_0 = 20$ m

Substituting the preceding information in Equation IP5.1.2 yields the following:

$$19 = \frac{Q}{2\pi KH} \ln\left[\frac{1000}{2000}\right] + 20 \quad \text{(IP5.1.3)}$$

From Equation IP5.1.3, the following can be obtained:

$$\frac{Q}{2\pi KH} = \frac{19 - 20}{\ln(0.5)} = 1.4427 \text{ m} \quad \text{(IP5.1.4)}$$

At point P(1000, 500), the following data can be easily obtained:

$r_1 = 707.107$ m

$r_2 = 1581.14$ m

$\phi_0 = 20$ m

Thus, substituting the preceding values in Equation IP5.1.2 yields the following:

$$\phi_P = 1.4427 \ln\left[\frac{707.107}{1581.13}\right] + 20 = 18.84 \text{ m}.$$

Part (*b*): The steady-state discharge can be obtained from Equation IP5.1.4 as follows:

$$Q = 1.4427\ \text{m} \times 2\pi \times K \times H = 1.4427\ \text{m} \times 2\pi \times 0.00014\ \text{m/s} \times 8\ \text{m}$$
$$= 0.010\ \text{m}^3/\text{s}.$$

5.2 A fully penetrating well of radius r_w is situated at a distance d from a perennial river, as shown in the sketch in section a. Groundwater is pumped from the well at a constant rate Q. Assuming the aquifer remains confined do the following:

(a) Develop an expression for time T taken by a nondispersive contaminant to travel from the disposal site d_c to the well. Assume the confined aquifer is homogeneously isotropic with coefficient of permeability K, porosity n, and depth H. Ignore the radius r_w in comparison with d and d_c.

(b) Find the travel time T in years for the following data: $d = 1000$ m, $d_c = 5000$ m, $K = 0.0001$ m/s, $H = 20$ m, $Q = 0.1$ m³/s, and $n = 0.25$.

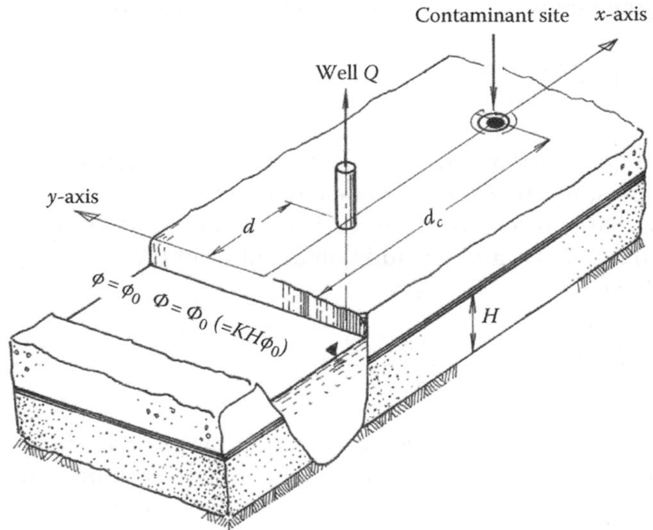

Part (*a*): The discharge potential $\Phi(x, y)$ in this case is given by the following equation:

$$\Phi(x,y) = \frac{Q}{4\pi}\left[\ln r_1^2 - \ln r_2^2\right] + \Phi_0 \tag{IP5.2.1}$$

where r_1 and r_2 denote the radial distances from the well and its image to a movable point P(x, y), respectively. The reader is reminded that the well in this case is situated at a distance d from the origin of the x, y, z-coordinate *system* on the positive x-axis. Thus, the radial distances can be expressed as follows:

$$r_1^2 = (x-d)^2 + y^2 \tag{IP5.2.2a}$$

$$r_2^2 = (x+d)^2 + y^2 \tag{IP5.2.2b}$$

The discharge $Q_x(x, y)$ per width normal to the x-axis through the entire depth of aquifer can be obtained by a partial differentiation of $\Phi(x, y)$ with respect to x. Thus, from Equations IP5.2.1 and IP5.2.2, we obtain the following expression

$$Q_x(x,y) = -\frac{\partial \Phi}{\partial x} = -\frac{Q}{4\pi}\left[\frac{2(x-d)}{(x-d)^2+y^2} - \frac{2(x+d)}{(x+d)^2+y^2}\right] \tag{IP5.2.3}$$

for discharge Q_x at any point P(x, y). In order to obtain the discharge along the x-axis, we set $y=0$ in the preceding equation to obtain

$$Q_x(x,0) = -\frac{Q}{\pi}\left[\frac{d}{x^2-d^2}\right] \tag{IP5.2.4}$$

From Equation IP5.2.4, it is evident that Q_x is negative for points on the x-axis, meeting the inequality requirement $x>d$. In other words, the discharge is toward the well. The x-component, q_x, of the specific discharge vector can be readily obtained from the preceding equation as follows:

$$q_x(x,0) = \frac{Q(x,0)}{H} = -\frac{Q}{\pi H}\left[\frac{d}{x^2-d^2}\right] \tag{IP5.2.5}$$

Equation IP5.2.5 describes the variation of q_x along the x-axis. The term *nondispersive* contaminant implies that the contaminant does not diffuse or disperse during convection by the flow. Thus, the contaminant travels with the average seepage velocity, v, through the aquifer. If $x=x(t)$ denotes the location of a parcel of contaminant on the x-axis at any time t, then the velocity of this parcel is given by

$$v(t) = \frac{dx}{dt}$$

Since the average seepage velocity, v, and the specific discharge, q, are related by equation $nv=q$ (see Chapter 2, Equation 2.50), it follows from Equation IP5.2.5 that the following statement is true:

$$n\frac{dx}{dt} = q_x(x,0) = -\frac{Q}{\pi H}\left[\frac{d}{x^2 - d^2}\right] \tag{IP5.2.6a}$$

where n denotes the areal (and also the volumetric) porosity. Thus, the migration of a nondispersive contaminant along the x-axis is governed by the following differential equation:

$$n\frac{dx}{dt} = -\frac{Q}{\pi H}\left[\frac{d}{x^2 - d^2}\right] \tag{IP5.2.6b}$$

The preceding differential equation is separable, for n, Q, d, and H are given (or known) constants. This equation can be integrated to obtain

$$n\int_{d_c}^{d}\left(x^2 - d^2\right)dx = -\frac{Qd}{\pi H}\int_{0}^{T}dt \tag{IP5.2.6c}$$

where T denotes the time taken by the nondispersive contaminant to travel from the disposal site, $x = d_c$, to the well, $x = d$. The solution of the preceding differential equation can be written in a dimensionless form as follows:

$$n\left[\frac{1}{3}\left(\frac{d_c}{d}\right)^3 - \left(\frac{d_c}{d}\right) + \frac{2}{3}\right] = \frac{T}{\pi H d^2 / Q} \tag{IP5.2.7}$$

The denominator on the right-hand side has the dimension of time, and for a given problem, this time is a unique constant. Let this constant be denoted by

$$\tau \equiv \frac{\pi H d^2}{Q}$$

It is possible to give this constant a physical meaning, for instance, it represents the time to empty a full cylindrical tank of radius d and height H while emptying the tank at a constant rate Q. A graph of Equation IP5.2.7 is illustrated in the following figure. In this graph, the abscissa denotes the normalized coordinate x/d, and the ordinate, the dimensionless time, T/τ. The curve parameter represents the porosity, n, of the aquifer. The curve with the curve parameter $n = 1$ is not physically realistic for aquifers and is only shown for completeness. Likewise, the curve for $n = 0$ is a degenerate case. Despite these comments, the following observations are pertinent:

(i) The travel time does not depend on the coefficient of permeability K, as long as the well is pumped at a constant rate of Q and the aquifer remains confined.
(ii) Reducing the porosity n of the aquifer (for instance, by compaction) only aggravates the travel time of the nondispersive contaminant. In other words, less porous aquifers transmit a contaminant at a faster rate, provided the well discharge is kept at the same constant rate Q and the aquifer remains confined.

At first glance, the preceding comments seem to defy common sense, but a moment of reflection reconciles the apparent contradiction in thought. The basic physical concepts involved in arriving at these conclusions are the notion of continuity of flow and the notion of average seepage velocity. As long as the flow field is continuous, the travel time decreases with a decrease in porosity.

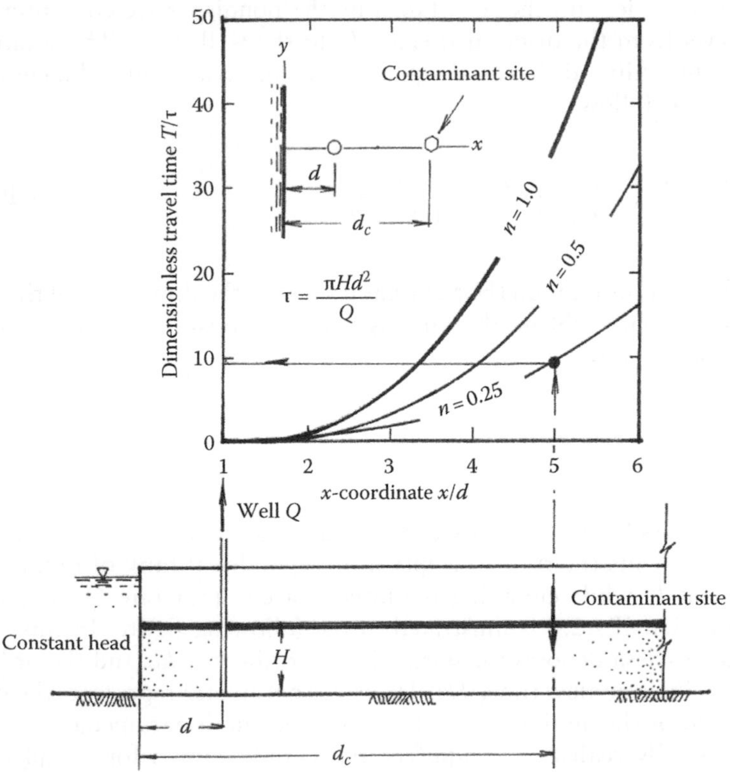

Part (*b*): For this problem, the value of K is not germane. Substituting the appropriate values for other variables in Equation IP5.2.7 yields the required answer for T as follows:

The normalizing time,

$$\tau = \frac{\pi H d^2}{Q} = \frac{\pi \times 20\,\text{m} \times (1000)^2\,\text{m}^2}{0.1[\text{m}^3/\text{s}]}$$

$$= 628{,}318{,}530.7\,\text{s} \times \left[\frac{1\,\text{day}}{86{,}400\,\text{s}}\right] = 7{,}272.2\,\text{day}$$

For this problem, the disposal site is located at the normalized distance, $d_c/d = 5$. Thus, substituting this value in the left-hand side of Equation IP5.2.7, along with a value of porosity, yields the following result:

$$n\left[\frac{1}{3}\left(\frac{d_c}{d}\right)^3 - \left(\frac{d_c}{d}\right) + \frac{2}{3}\right] = 0.25\left[\frac{1}{3}(5)^3 - 5 - \frac{2}{3}\right] = 9.33333$$

Finally, the travel time is obtained as follows:

$$T = 9.33333 \times \tau = 9.33333 \times 7272.2\ \text{day}$$

$$T = 9.33333 \times 7272.2\ \text{day} \times \left[\frac{1\,\text{year}}{365\,\text{day}}\right] = 186\ \text{year}$$

5.3 A fully penetrating well with eccentricity ratio $\delta/R = 0.5$ is located in a circular island of diameter 4000 m. The island is surrounded by a freshwater lake of depth 20 m. The average depth of the confined aquifer is estimated to be $H = 10$ m (see the following figure).

(a) Find the maximum steady discharge of the well so that the aquifer remains confined, if the radius of the well $r_w = 0.25$ m, and the estimated coefficient of isotropic permeability $K = 0.00015$ m/s.

(b) Repeat the same problem if the well is centrally located, that is, $\delta/R = 0.0$. Compare the results in the two cases.

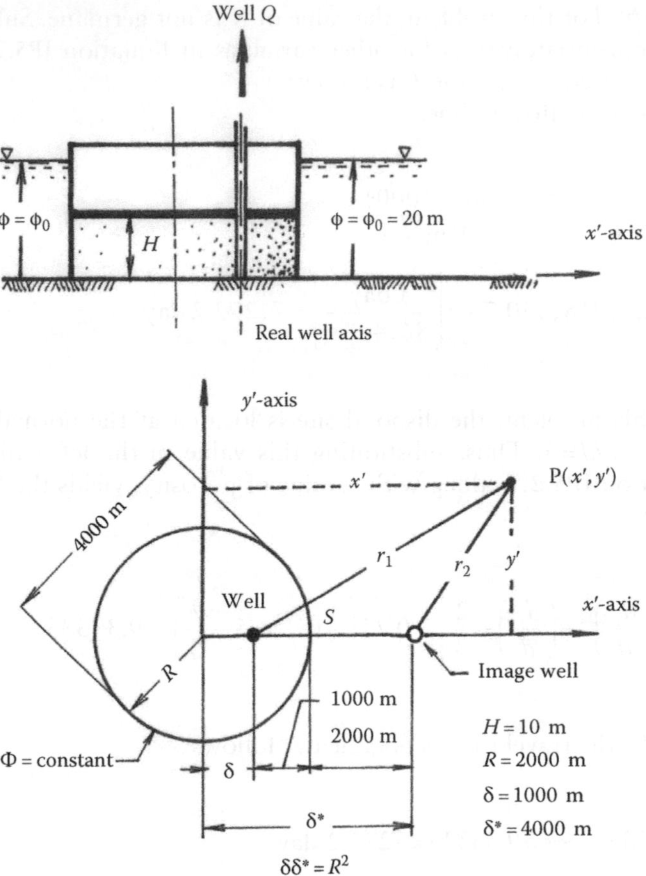

Part (*a*): The discharge potential Φ at any point $P(x', y')$ is given by the following equation:

$$\Phi = \frac{Q}{4\pi}\left[\ln\frac{r_1^2}{r_2^2}\right] + C \tag{IP5.3.1}$$

where

r_1 and r_2 depend on the location of the movable point $P(x', y')$
C is an arbitrary constant

If desired, the radial distances can be expressed in terms of x' and y' *coordinates* by the following equations:

$$r_1^2 = (x' - \delta)^2 + y'^2 \tag{IP5.3.2a}$$

$$r_2^2 = \left(x' - \delta^*\right)^2 + y'^2 \tag{IP5.3.2b}$$

$$\delta^* = \frac{R^2}{\delta} \tag{IP5.3.2c}$$

Determination of arbitrary constant C: The arbitrary constant in Equation IP5.3.1 can be determined by invoking the condition on the circumference of the island, that is, $\Phi = \Phi_0 = KH\phi_0 = 0.00015$ m/s × 10 m × 20 m = 0.030 m³/s. Since $\Phi = \Phi_0$ is constant at every point on the circumference, this must also be true for a particular point S on the circumference. Thus, at point S, the following statements are true:

$$\Phi = \Phi_0 (\text{a known constant}) \tag{IP5.3.3a}$$

$$r_1^2 = (R - \delta)^2 \tag{IP5.3.3b}$$

$$r_2^2 = \left(R - \delta^*\right)^2 \tag{IP5.3.3c}$$

Substituting the preceding values in Equation IP5.3.1 and replacing δ^* by R^2/δ yield the following expression for the arbitrary constant:

$$C = \Phi_0 - \frac{Q}{4\pi} \ln \frac{\delta^2}{R^2} \tag{IP5.3.4}$$

Thus, combining Equation IP5.3.4 with Equation IP5.3.1 yields the potential at any point in terms of the known parameters of the problem, as follows:

$$\Phi = \frac{Q}{2\pi}\left[\ln \frac{r_1}{r_2} + \ln \frac{R}{\delta}\right] + \Phi_0 \tag{IP5.3.5}$$

For the aquifer to remain confined, the piezometric head at the well must not fall below 10 m. It is, therefore, true that the preceding equation reduces to the following at the circumference of the well:

$$\Phi_w = \frac{Q}{2\pi}\left[\ln \frac{r_w}{r_2} + \ln \frac{R}{\delta}\right] + \Phi_0 \tag{IP5.3.6}$$

where

$$\Phi_w = KH\phi_w = 0.00015 \text{ m/s} \times 10 \text{ m} \times 10 \text{ m} = 0.015 \text{ m}^3/\text{s}$$
$$r_w = 0.25 \text{ m}$$

$\delta = 1000$ m
$R = 2000$ m
$r_2 = 3000\,\text{m} - 0.25\,\text{m} \cong 3000\,\text{m}.$

Substituting the preceding values in Equation IP5.3.6 yields the following:

$$0.015\,\text{m}^3/\text{s} = \frac{Q}{2\pi}\left[\ln\frac{0.25}{3000} + \ln\frac{2000}{1000}\right] + 0.030\,\text{m}^3/\text{s}$$

from which the unknown Q can be obtained as

$$Q = 0.0108\ \text{m}^3/\text{s} \qquad \text{(IP5.3.7)}$$

Part (*b*): In this case, there is complete axial symmetry. Hence, the following equation can be used:

$$\Phi(r) = \frac{Q}{2\pi}\left[\frac{r}{r_0}\right] + \Phi_0$$

Substituting $r = r_w = 0.25\,\text{m}$ and $r_0 = R = 2000\,\text{m}$ in the preceding equation yields the following:

$$0.015\,\text{m}^3/\text{s} = \frac{Q}{2\pi}\left[\ln\frac{0.250}{2000}\right] + 0.030\,\text{m}^3/\text{s}$$

From the preceding equation, Q is obtained as

$$Q = 0.0105\ \text{m}^3/\text{s}. \qquad \text{(IP5.3.8)}$$

Now, comparing this value of Q with the previous value (Equation IP5.3.7), it is obvious that the two values differ by less than 5% despite the fact that the eccentricity ratio in the two cases differ by as much as 50%. This conclusion is in conformity with the previous findings (see page 80 of Todd [1959]).

5.5 EXERCISES

5.1 For Illustrative Problem 5.1, determine the contour of the piezometric surface that passes through the observation well B. The contour of the piezometric surface is a curve along which the piezometric level remains the same. (*Hint*: In this case, we know the contours are circular curves. Find the point closest to the constant-head boundary where the piezometric head is the same as $\phi_B = 19$ m. Then find the diameter and the radius of the circular contour.)

5.2 For the case discussed in Illustrative Problem 5.1, prove the aquifer remains confined, if the well radius $r_w = 0.25$ m.

5.3 Prove that the definite integral

$$Q_{total} = 2\int_0^{\infty} Q_x dy$$

represents the steady (not changing with time) well discharge Q, if (see Equation 5.7a)

$$Q_x = -\frac{Q}{\pi d}\left[\frac{1}{1+(y/d)^2}\right]$$

where Q and d are constants.

SUGGESTED READINGS

Related to the mechanics of groundwater flow:

Bear, J. 1988. *Dynamics of Fluids in Porous Media*, Dover Publications, Inc., New York.

Hermance, J. F. 1999. *A Mathematical Primer on Groundwater Flow*, Prentice Hall, Upper Saddle River, NJ.

Strack, O. D. L. 1989. *Groundwater Mechanics*, Prentice Hall, Inc., Englewood Cliffs, NJ.

Todd, D. K. 1959. *Ground Water Hydrology*, John Wiley & Sons, Inc., New York.

Mathematics applied to the Laplace equation:

Churchill, R. V. and J. W. Brown. 1990. *Complex Variables and Applications*, 5th edn., McGraw-Hill Inc., New York.

Kellogg, O. D. 1953. *Foundations of Potential Theory*, Dover Publications Inc., New York.

Rothe, R., Ollendorff, F., and K. Pohlhausen (Eds.). 1961. *Theory of Functions as Applied to Engineering Problems*, Dover edition, Dover Publications Inc., New York.

Sneddon, I. N. 1957. *Elements of Partial Differential Equations*, McGraw-Hill Book Company, Inc., New York.

Chapter 6

Flow net

In this chapter, we intend providing the theoretical basis and the implementation of the graphical method, commonly called the method of *flow net*, for estimating seepage and pore (uplift) pressure in a two-dimensional groundwater flow. Although the graphical method can be applied to two-dimensional flows in a horizontal plane, it is mostly applied to two-dimensional flows in a vertical plane, under hydraulic structures, especially by civil engineers. Through most of this chapter, we shall assume that the vertical plane is spanned by x, y *coordinates* of a rectangular Cartesian coordinate system, where y-axis points vertically upward. We shall first describe the method of flow net for an isotropic, homogeneous aquifer; later, we shall extend the method to include anisotropic, heterogeneous media. *We must, however, state that the graphical method of flow net is not applicable to three-dimensional flow fields.*

6.1 ISOTROPIC CASE

In the case of steady two-dimensional (planar) groundwater flows in homogeneous, isotropic media, we know from Darcy's law that the components of the specific discharge vector $q=(q_x, q_y)$ at any point (x, y) can be derived from a potential function as shown in the following:

$$q_x(x,y) = -\frac{\partial \Phi(x,y)}{\partial x} \tag{6.1a}$$

$$q_y(x,y) = -\frac{\partial \Phi(x,y)}{\partial y} \tag{6.1b}$$

where the potential function $\Phi(x, y) = K\phi(x, y) + C$. Here, the symbols K and ϕ have their usual meanings: the isotropic coefficient of permeability and the piezometric head at any point (x, y) in the planar flow field. We can, for convenience, set the arbitrary constant, C, to zero. This is tantamount to setting the arbitrary datum for Φ and ϕ at the same level. In order for Φ to be called a potential function, Φ must also satisfy the Laplace equation. Since by hypothesis the flow is in a two-dimensional, steady state, it must satisfy the continuity equation:

$$\frac{\partial q_x}{\partial x} + \frac{\partial q_y}{\partial y} = 0 \tag{6.2}$$

Although it is not shown explicitly, it is understood that q_x and q_y are functions of two independent coordinates x and y. Now, by substituting the expressions of q_x and q_y from Equation 6.1 into Equation 6.2, we readily see that Φ is indeed a harmonic function. It is, therefore, properly called the potential function. A flow field in which the velocity components are derivable from a scalar potential and which also satisfies the continuity equation is called a *potential flow* in fluid mechanics. Thus, a steady two-dimensional groundwater flow in an isotropic, homogeneous aquifer can also be called a *potential flow*.

Now, for a planar flow, $\Phi(x, y)$ is a scalar function of two independent spatial coordinates x, y. Thus, $\Phi(x, y) = \text{constant}$ represents a curve in the x, y-plane (Figure 6.1). This curve is also called an equipotential curve, because along this curve potential remains the same. Thus, along this curve, the differential $d\Phi = 0$. We know from calculus that the exact differential of $\Phi(x, y)$ is given by the following equation:

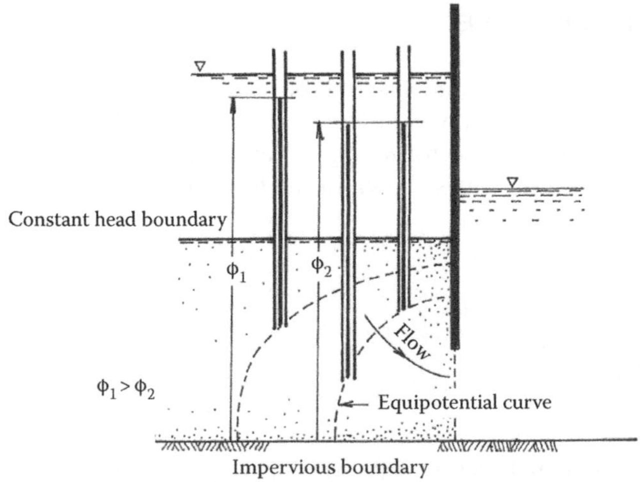

Figure 6.1 Definition sketch for an equipotential curve.

$$d\Phi = \frac{\partial \Phi}{\partial x}dx + \frac{\partial \Phi}{\partial y}dy \tag{6.3a}$$

Thus, along $\Phi(x, y) = $constant curves, the following must be true:

$$d\Phi = \frac{\partial \Phi}{\partial x}dx + \frac{\partial \Phi}{\partial y}dy = -\left(q_x dx + q_y dy\right) = 0 \tag{6.3b}$$

where the second equality follows from the fact that the flow is a potential flow. The last equation yields the following differential equation for $\Phi(x, y) = $constant curves:

$$\frac{dy}{dx}\Big|_{\Phi=const} = -\frac{q_x}{q_y} \tag{6.4}$$

Let us, now, consider a steady two-dimensional flow field in a region devoid of any point source or sink, as shown in Figure 6.2. Since the flow field is steady, the discharge, per width normal to the plane of paper, across curve I must be the same as the discharge across curve II. In other words, the discharge depends on the locations of point s_1 and s_2 only and not on the geometry of the curve, *per se*. Thus, the following line integral

$$\Delta Q = \int_{s1}^{s2} \boldsymbol{q}\cdot\boldsymbol{n}\,ds = \int_{s1}^{s2}\left(q_x n_x + q_y n_y\right)ds \tag{6.5}$$

must be independent of the actual path of integration. This is only possible if the expression $\boldsymbol{q}\cdot\boldsymbol{n}\,ds$ is an exact differential of some function of x and y. Let this function be denoted by $\Psi(x, y)$. Thus,

$$\Delta Q = \int_{\Psi_1}^{\Psi_2} d\Psi = \int_{s1}^{s2}\left(q_x n_x + q_y n_y\right)ds \tag{6.6a}$$

and

$$\Delta Q = \Psi_2 - \Psi_1 \tag{6.6b}$$

The preceding argument is predicated on the premise that the flow field is *steady and planar*. It does not assume that the velocity field is derivable from a potential. Thus, for all *planar flow fields* (whether potential or not), *which happen to be in steady-state conditions, there exist scalar functions* $\Psi(x, y)$, such that the preceding two equations are satisfied. This reasoning

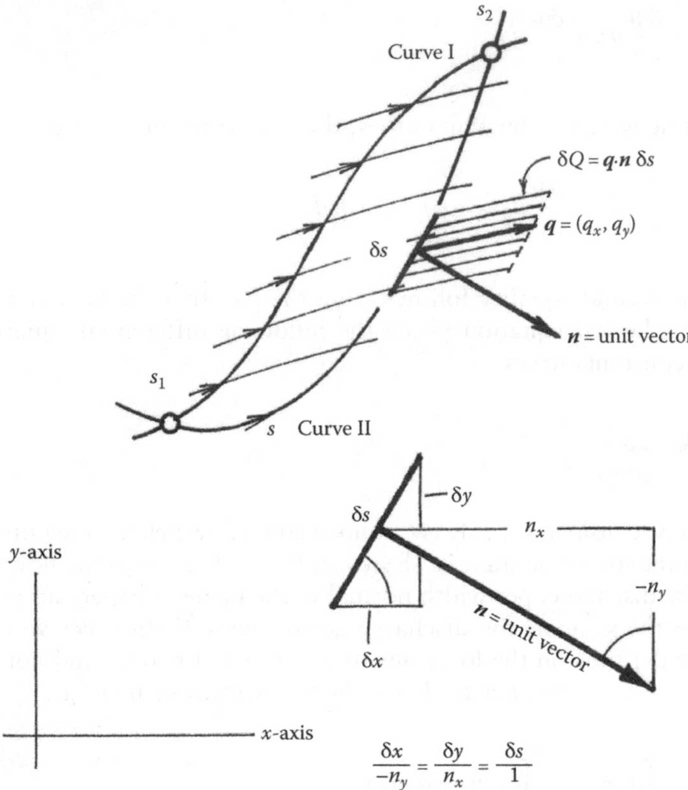

Figure 6.2 Two-dimensional steady flow and stream function.

only establishes the existence of $\Psi(x, y)$ in *planar, steady flow fields* but reveals little on the physical nature of such functions, other than that envisioned earlier. To further explore the physical meaning and other mathematical manifestations of such functions, we proceed as follows.

Equating the expressions inside the integral signs of Equation 6.6a, we get

$$d\Psi = \left(q_x n_x + q_y n_y\right) ds \tag{6.7}$$

where (see Figure 6.2)

$$n_x = \lim_{\delta s \to 0} \frac{\delta y}{\delta s} = \frac{dy}{ds} \tag{6.8a}$$

$$n_y = \lim_{\delta s \to 0} -\frac{\delta x}{\delta s} = -\frac{dx}{ds} \tag{6.8b}$$

or

$$d\Psi = \left(q_x dy - q_y dx\right) \tag{6.9a}$$

From calculus, the exact differential of a function $\Psi(x, y)$ is given by the following equation:

$$d\Psi = \frac{\partial \Psi}{\partial x}dx + \frac{\partial \Psi}{\partial y}dy \tag{6.9b}$$

Thus, from Equations 6.9a and 6.9b, we obtain the following:

$$d\Psi = \left(q_x dy - q_y dx\right) = \frac{\partial \Psi}{\partial x}dx + \frac{\partial \Psi}{\partial y}dy \tag{6.9c}$$

or

$$\left(\frac{\partial \Psi}{\partial x} + q_y\right)dx + \left(\frac{\partial \Psi}{\partial y} - q_x\right)dy = 0 \tag{6.9d}$$

Since *dx* and *dy* are independently arbitrary (i.e., one can change arbitrarily *dx* or *dy* without simultaneously changing the other), it follows from the last equation that the following two equations must be true:

$$q_x = +\frac{\partial \Psi}{\partial y} \tag{6.10a}$$

$$q_y = -\frac{\partial \Psi}{\partial x} \tag{6.10b}$$

The last two equations show that the components of specific discharge vector q are also related to the partial derivatives of $\Psi(x, y)$.

Now, we introduce the notion of a *streamline* in groundwater flow, parallel to a similar concept in fluid mechanics. A streamline is an imaginary line (in general, a curved line) in a groundwater flow field such that at every point on this line the specific discharge vector q is tangential to the line. Although the concept of a streamline is valid for a three-dimensional flow field, we restrict it to two-dimensional planar flows only. Thus, the tangent at any point of a streamline must satisfy the following differential equation (see Figure 6.3):

$$\frac{dy}{dx}\Big|_{streamline} = \frac{q_y}{q_x} \tag{6.11}$$

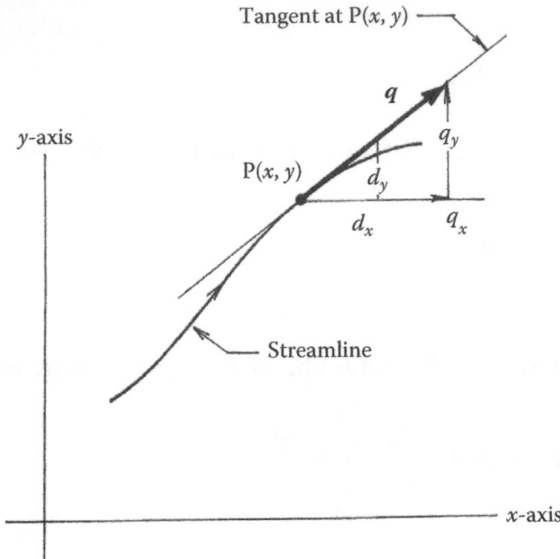

Figure 6.3 Definition sketch for the derivation of streamlines.

Since $\Psi(x, y)$ is a function of two independent variables, x and y, $\Psi = constant$ represents a curve in x, y-plane. Along this curve, $d\Psi = 0$. Hence, setting the differential $d\Psi$ in Equation 6.9a equal to zero gives the following differential equation for all $\Psi = constant$ curves:

$$\frac{dy}{dx}\Big|_{\Psi=const} = \frac{q_y}{q_x} \tag{6.12}$$

Comparing Equation 6.12 with Equation 6.11, it becomes clear that $\Psi = constant$ curves are the same as streamlines, for their differential equations are the same. It is for this reason that the scalar function $\Psi(x, y)$ is called the *stream function*, because for arbitrary values of constant C, $\Psi(x, y) = C$ represents a family of streamlines in the flow field.

At a give point (x, y), the slope of $\Phi(x, y) =$ constant curve is given by Equation 6.4 and, at the same point, the slope of $\Psi(x, y) =$ constant curve is given by Equation 6.12. Thus multiplying the two slopes, we obtain the following:

$$\frac{dy}{dx}\Big|_{\Psi=const} \times \frac{dy}{dx}\Big|_{\Phi=const} = \frac{q_y}{q_x} \times \left(-\frac{q_x}{q_y} \right) = -1 \tag{6.13}$$

This implies that the constant-Ψ and constant-Φ curves intersect each other at right angles. Thus, these curves constitute the so-called *orthogonal trajectories*. It is worth stressing at this point that the orthogonal intersection of these curves is only ensured in steady, planar groundwater flows in homogeneous, isotropic aquifers. If the medium is not *homogeneously isotropic*, the specific discharge vector $q(q_x, q_y)$ is not derivable from a scalar potential function Φ. Likewise, if the flow is *not steady and planar*, the stream function Ψ does not exist. In other words, if either Φ or Ψ does not exist, it is mute to talk about orthogonal trajectories.

Finally, if both potential (Φ) and stream functions (Ψ) exist, then each family contains infinitely many members (or curves), because there are infinitely many solutions to a differential equation, such as Equation 6.4 or 6.12. However, a finite number of these orthogonal trajectories can be obtained graphically. Such a finite number of orthogonal trajectories constitute the so-called *flow net* for the given problem.

It is sometimes more convenient to work with the piezometric head ϕ, instead of Φ. In this regard, it is worth stressing that along an equipotential curve both ϕ and Φ ($=K\phi$) remain constant. Thus, $\phi=$constant curves also intersect streamlines orthogonally. It is therefore evident that constant piezometric head lines and streamlines also constitute the orthogonal trajectories. A finite number of these orthogonal trajectories can be used to find flow net for a given problem. The guidelines for obtaining the graphical solution are further discussed using a concrete example.

6.1.1 Example of flow-net construction and analysis

An example of flow net is shown in Figure 6.4. In this figure, AB and CD represent that part of the boundaries where the piezometric head ϕ is known. Because of the presence of large bodies of water in a quasi-static state, the piezometric head along these boundaries is assumed constant. For instance, on AB, the piezometric head $\phi=18$ m and on the boundary CD, $\phi=2$ m. For the purpose of assigning numerical values to ϕ, the arbitrary datum is assumed at the same level as that of AB. A part of the top segment of the flow region, denoted by 1-2-3-4-5, represents a streamline along which Ψ remains constant. It is assumed, for convenience, that the contact line between the structure and the aquifer represents a streamline because of the substantial difference between the coefficients of permeability of the aquifer and the concrete structure. Likewise, the lower boundary, denoted by 6-7, represents another streamline, which represents another $\Psi=$constant boundary.

In the foregoing theoretical development, we have seen that the $\Psi=$ constant curves (streamlines) intersect $\phi=$constant curves (equipotential lines) at right angles. During the graphical construction of flow net, this fact is kept in mind, while sketching the $\Psi=$constant (streamlines) and

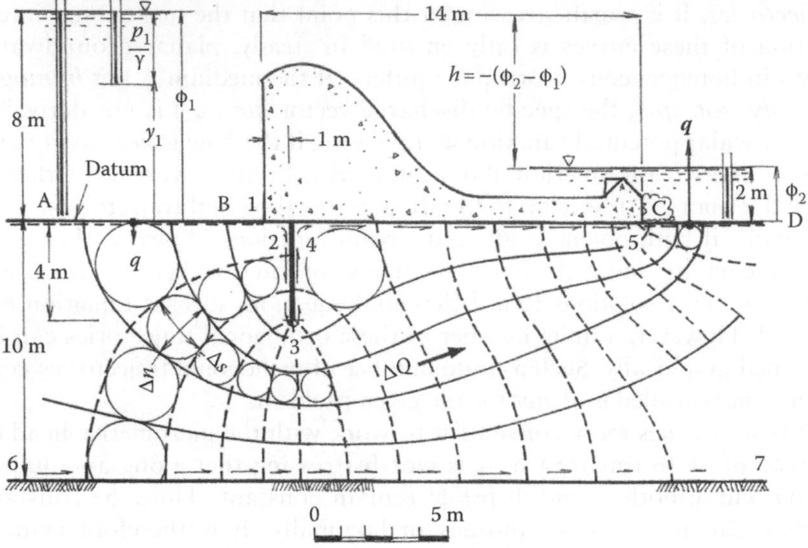

Figure 6.4 An example of flow net in a homogeneous, isotropic aquifer.

the ϕ = constant (equipotential) lines to cover the main portion of the flow domain. To start the drafting process, two or three streamlines are tentatively sketched, making sure they intersect at right angles to the known equipotential lines at AB and CD. Then, two or three equipotential lines are tentatively sketched, making sure that they intersect the known streamlines, such as 1-2-3-4-5 and 6-7, at right angles. In addition, they should also intersect the tentatively drawn streamlines close to right angles. If the intersections between the tentatively drawn streamlines and equipotential lines are far from being at right angles, they are continually adjusted until they approximately satisfy this condition. During this trial-and-error effort, the intersections of streamlines at the known boundaries, such as AB and CD, are kept at perfect right angles. Likewise, the intersections of sketched equipotential lines with the known streamlines, such as 1-2-3-4-5 and 6-7, are maintained at right angles. During this process, the experience of the draftsman and familiarity with previously well-drawn flow nets—especially with similar flow domain and specified boundary conditions—are extremely important. Although drawing good flow nets is an art, this skill can be developed by an initiated beginner with enough practice.

Once the task of drawing the initial streamlines and equipotential lines is satisfactorily accomplished, the task of filling in the remaining flow domain with additional streamlines and equipotential lines begins.

With regard to the construction of flow net, the following guidelines are helpful:

1. The streamlines and equipotential lines should be smooth curves and intersect each other orthogonally. Realizing that the top streamline (1-2-3-4-5) is very convoluted and the lowest streamline (6-7) is rather straight forward, it is helpful to perceive a gradual transition from the top to the lowest streamline while sketching intermediate streamlines.
2. Each flow channel, contained between two consecutive streamlines, must preferably carry equal amount of flow.
3. Although the theory demands only orthogonal intersection between the streamline and the equipotential line, it puts no restrictions on the aspect ratio (ratio between length and width) of the individual cell of the flow net—whether it should represent a curvilinear square or a curvilinear rectangle.
4. However, from a practical point of view, if we plan to use the flow net to estimate seepage through the aquifer, or predict the uplift pressure due to porewater, it is incumbent upon the analyst to draw the flow net with individual cells consisting of curvilinear squares. Only in such a case, different flow channels carry equal amount of flow, and the consecutive equipotential lines represent equal drop in the piezometric head.
5. Although not necessary, it is however convenient to maintain an aspect ratio of one. In other words, each cell of the flow net should represent a curvilinear square.

In the construction of flow net exemplified in Figure 6.4, the aforementioned guidelines have been used with reasonable adherence. In such a flow net, the individual cell can be inscribed by a circle. A typical inscribed circle with curvilinear diameters labeled as Δs *and* Δn is also shown in the figure. The inscription of a circle, thus, demonstrates that the individual cell represents a curvilinear square instead of a curvilinear rectangle. It is this cell that approaches a perfect square as its diameter reduces *ad infinitum*.

The quantitative application of the flow net to the estimation of seepage and uplift pressure is described in *Illustrative Problem 6.1*.

6.2 ANISOTROPIC CASE

We assume that a given medium is homogeneously anisotropic with principal values of permeability K'_{xx} and K'_{yy} in the principal directions, which are taken along x'-axis and y'-axis, respectively. In this medium, according to Darcy's law, the components q'_x and q'_y of the specific discharge vector, along the principal directions, at any point (x', y'), are given by the following equation:

$$q'_x(x',y') = -K'_{xx}\frac{\partial\phi}{\partial x'} \quad (6.14a)$$

$$q'_y(x',y') = -K'_{yy}\frac{\partial\phi}{\partial y'} \quad (6.14b)$$

If the flow is two dimensional in a steady-state condition, the following continuity equation must be satisfied:

$$\frac{\partial q'_x}{\partial x'} + \frac{\partial q'_y}{\partial y'} = 0 \quad (6.15)$$

It may be emphasized that the continuity equation is valid for both isotropic and anisotropic media, as long as the fluid and the porous medium remain incompressible, and the flow field is in a steady-state condition. Substituting the components of the specific discharge vector from Equation 6.14a and 6.14b into Equation 6.15 yields the following:

$$K'_{xx}\frac{\partial^2\phi}{\partial x'^2} + K'_{yy}\frac{\partial^2\phi}{\partial y'^2} = 0 \quad (6.16)$$

It is evident that ϕ is neither a potential function nor it satisfies the Laplace equation, for Equation 6.16 does not lead to the Laplace equation as long as $K'_{xx} \neq K'_{yy}$. In other words, the scalar function $\phi(x', y')$ does not satisfy the Laplace equation in anisotropic media. It is therefore not a potential function and the resulting flow field (Equations 6.14a and 6.14b) is not a potential flow field. Thus, in anisotropic media, constant-ϕ curves do not intersect constant-Ψ curves at right angles.

At this juncture, we ask whether it is possible to transform the physical flow region into a fictitious flow region where Equation 6.16 indeed reduces to the Laplace equation. The short answer is yes, and it can be easily justified by choosing the following transformation:

$$x = x' \quad (6.17a)$$

$$y = \beta y' \quad (6.17b)$$

In the aforementioned equations, x y *coordinates* span the new fictitious plane, β is a magnification factor (if $\beta > 1$), x' y' *coordinates* span the anisotropic physical plane, and the two sets of coordinate axes—x'-axis, y'-axis

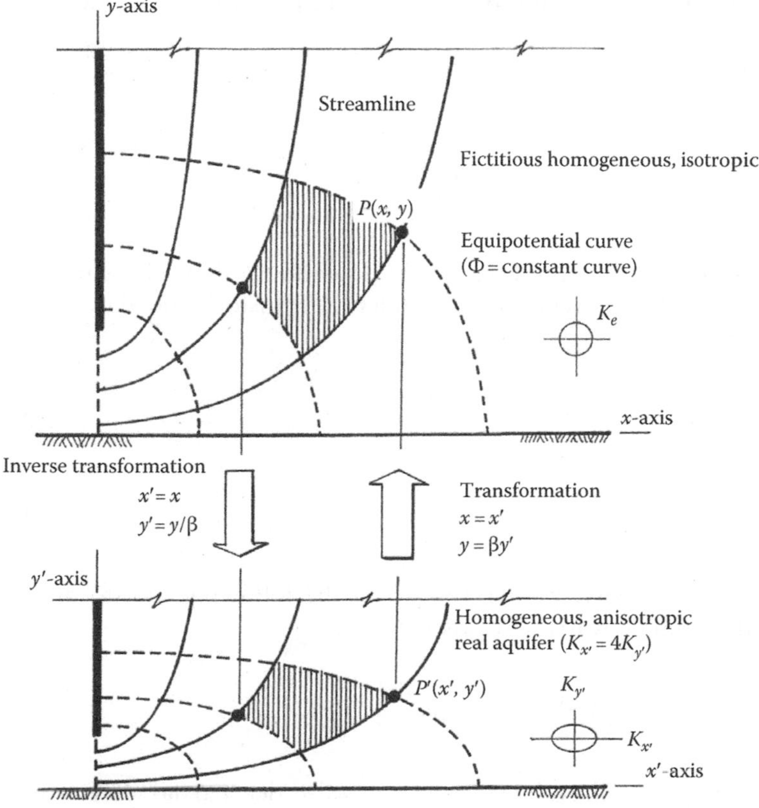

Figure 6.5 Mapping of an anisotropic aquifer to a fictitious isotropic flow domain, using transformation Equation 6.17.

and x-axis, y-axis—are, respectively, parallel under the aforementioned transformation.

This transformation maps an antecedent point $P'(x', y')$ onto an image (transformed) point $P(x, y)$ (Figure 6.5). Thus, in general, under this transformation, the x *coordinate* of P becomes some function of x', y'; that is, $x = x(x', y')$, because image P depends on the location of its antecedent point P', defined by its coordinates (x', y'). Likewise, $y = y(x', y')$. Now let the scalar function ϕ be known as an explicit function of x, y *coordinates* of the transformed region; that is, $\phi = \phi(x, y)$. Then, the scalar function can also be written as an implicit function of x', y' coordinates as shown in the following:

$$\phi = \phi\left[x(x', y'), y(x', y')\right] \tag{6.18}$$

Using the rule for partial differentiation of implicit functions, we obtain

$$\frac{\partial \phi}{\partial x'} = \frac{\partial \phi}{\partial x}\frac{\partial x}{\partial x'} + \frac{\partial \phi}{\partial y}\frac{\partial y}{\partial x'} = \frac{\partial \phi}{\partial x} \tag{6.19a}$$

$$\frac{\partial \phi}{\partial y'} = \frac{\partial \phi}{\partial x}\frac{\partial x}{\partial y'} + \frac{\partial \phi}{\partial y}\frac{\partial y}{\partial y'} = \beta\frac{\partial \phi}{\partial y} \tag{6.19b}$$

In the second equality, in Equations 6.19a and 6.19b, we have made use of the fact that the following are true under the transformation law (Equations 6.17a and 6.17b)

$$\frac{\partial x}{\partial x'} = 1; \quad \frac{\partial y}{\partial x'} = 0 \tag{6.20a}$$

$$\frac{\partial x}{\partial y'} = 0; \quad \frac{\partial y}{\partial y'} = \beta \tag{6.20b}$$

In a similar manner, we can show that the following transformations of second partial derivatives are also true:

$$\frac{\partial^2 \phi}{\partial x'^2} = \frac{\partial^2 \phi}{\partial x^2} \quad \text{and} \quad \frac{\partial^2 \phi}{\partial y'^2} = \beta^2\frac{\partial^2 \phi}{\partial y^2} \tag{6.21}$$

Substituting the above-mentioned second-order partial derivatives into Equation 6.16 yields the following:

$$K'_{xx}\frac{\partial^2 \phi}{\partial x^2} + K'_{yy}\beta^2\frac{\partial^2 \phi}{\partial y^2} = 0 \tag{6.22}$$

If we choose $K'_{yy}\beta^2 = K'_{xx}$, Equation 6.22 reduces to the Laplace equation in the transformed region spanned by coordinates x and y. Thus, in a transformed space mapped by Equations 6.17a and 6.17b with $\beta = \sqrt{K'_{xx}/K'_{yy}}$, the scalar function ϕ becomes a harmonic function and the flow becomes the *planar potential flow*. In such a flow field, ϕ and Ψ curves constitute orthogonal trajectories, and, consequently, the construction of an orthogonal flow net becomes possible.

6.2.1 Equivalent permeability in transformed regions

In the foregoing, we found a way to transform a homogeneous, anisotropic region into a fictitious homogeneous, isotropic region in which Ψ and ϕ

curves intersect at right angles. These orthogonal trajectories can be further converted to skewed trajectories by the inverse transformation (Figure 6.5). Thus, the function ϕ and the corresponding pressure head (p/γ) can be found at any point (x', y') in the real anisotropic aquifer. However, the question that remains is: How do we find an equivalent isotropic coefficient of permeability, K_e, such that the seepage per width normal to the plane of paper through the transformed isotropic flow region remains the same as that through the real anisotropic aquifer? To answer this question, we proceed as follows:

In an attempt to answer this question, Figure 6.6 has been prepared. In Figure 6.6, a differential element $A'B'C'$ is shown, which under transformation becomes the differential element ABC as shown in the figure. *We require that the differential discharge, dQ', across the differential line segment, $B'C'$, should be the same as the differential discharge, dQ, across*

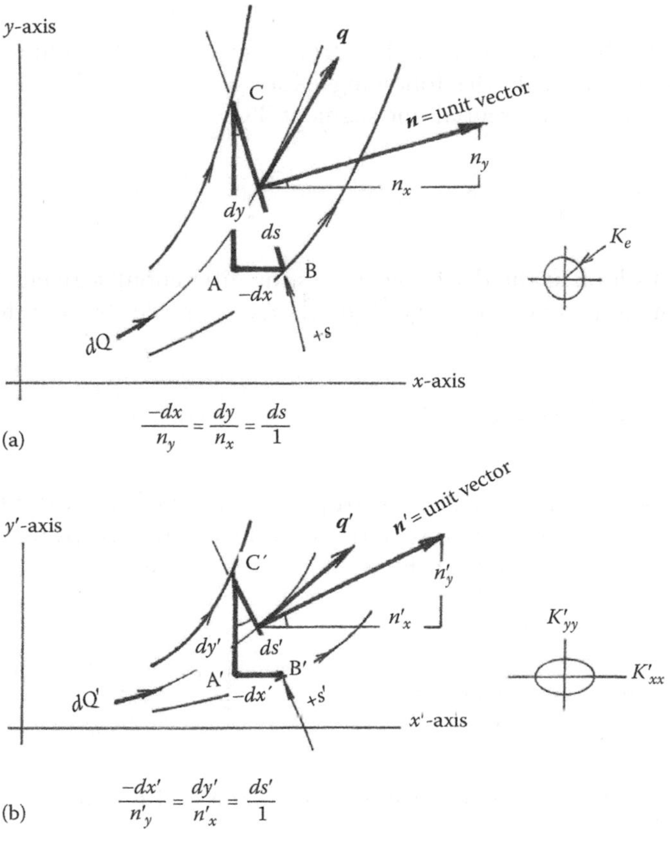

Figure 6.6 Definition sketch for equivalent isotropic coefficient of permeability: streamlines and differential element in (a) a fictitious isotropic aquifer and (b) a real anisotropic aquifer.

the transformed line segment, BC. The differential discharge across the differential line segment $B'C'$ is given by the following:

$$dQ' = q' \cdot n' ds' = \left(q'_x n'_x + q'_y n'_y\right) ds' = q'_x dy' - q'_y dx' \tag{6.23a}$$

Since q'_x and q'_y are the components of the specific discharge vector along the principle directions of the anisotropic aquifer, they can be written using Darcy's law as shown in the following:

$$q'_x = -K'_{xx} \frac{\partial \phi}{\partial x'} \tag{6.23b}$$

$$q'_y = -K'_{yy} \frac{\partial \phi}{\partial y'} \tag{6.23c}$$

Substituting the values of $q_{x'}$ and $q_{y'}$ from Equations 6.23b and 6.23c into Equation 6.23a yields the following expression for the differential discharge across the differential line segment B′C′:

$$dQ' = -K'_{xx} \frac{\partial \phi}{\partial x'} dy' + K'_{yy} \frac{\partial \phi}{\partial y'} dx' \tag{6.24}$$

Now, let us look at the discharge across the differential segment BC after transformation. This differential discharge is given by the following equation:

$$dQ = -K_e \frac{\partial \phi}{\partial x} dy + K_e \frac{\partial \phi}{\partial y} dx \tag{6.25}$$

where K_e denotes the equivalent isotropic coefficient of permeability (yet to be determined) of the fictitious isotropic aquifer. From the transformation law (Equations 6.17a and 6.17b), we note the following:

$$dy = \beta dy'; \quad dx = dx' \tag{6.26a}$$

$$\frac{\partial \phi}{\partial x} = \frac{\partial \phi}{\partial x'}; \quad \frac{\partial \phi}{\partial y} = \frac{1}{\beta} \frac{\partial \phi}{\partial y'} \tag{6.26b}$$

Substituting these values from Equations 6.26a and 6.26b into Equation 6.25 yields

$$dQ = -K_e \frac{\partial \phi}{\partial x'} \beta dy' + K_e \frac{1}{\beta} \frac{\partial \phi}{\partial y'} dx' \tag{6.27}$$

Since we require that the differential discharge, dQ', given in Equation 6.24, should be equal to the differential discharge, dQ, given in Equation 6.27, the right-hand sides of these two equations should also equal each other; that is,

$$-K'_{xx}\frac{\partial\phi}{\partial x'}dy' + K'_{yy}\frac{\partial\phi}{\partial y'}dx' = -K_e\frac{\partial\phi}{\partial x'}\beta dy' + K_e\frac{1}{\beta}\frac{\partial\phi}{\partial y'}dx' \tag{6.28}$$

or

$$(\beta K_e - K'_{xx})\frac{\partial\phi}{\partial x'}dy' - \left(K_e\frac{1}{\beta} - K'_{yy}\right)\frac{\partial\phi}{\partial y'}dx' = 0 \tag{6.29}$$

Since dx' and dy' are independently arbitrary, and both partial derivatives $\partial\phi/\partial x'$ and $\partial\phi/\partial y'$ cannot simultaneously vanish (if they do, the groundwater flow is stagnant), it follows that

$$K'_{xx} = \beta K_e \quad \text{and/or} \quad K'_{yy} = \frac{1}{\beta}K_e \tag{6.30}$$

Thus, if we demand that the discharge per width normal to the paper should be the same in both cases—the real anisotropic aquifer and the fictitious isotropic aquifer—then the equivalent coefficient of *isotropic permeability* of fictitious aquifer should be

$$K_e = \sqrt{K'_{xx}K'_{yy}} \tag{6.31}$$

Equation 6.31 finally gives the required result.

6.2.2 Example of flow-net construction in a homogeneous, anisotropic aquifer

The entire process of flow-net construction in a homogeneous, anisotropic flow domain requires three separate but complementary steps:

1. Mapping the anisotropic real flow domain into a fictitious isotropic domain, using the transformation indicated in Equations 6.17a and 6.17b. This aspect of the solution is illustrated in Figure 6.7.
2. Finding a flow net in a fictitious isotropic domain where the Laplace equation is valid; thus, the flow is potential and the construction of an orthogonal flow net is possible. This aspect of the solution is illustrated in Figure 6.8.

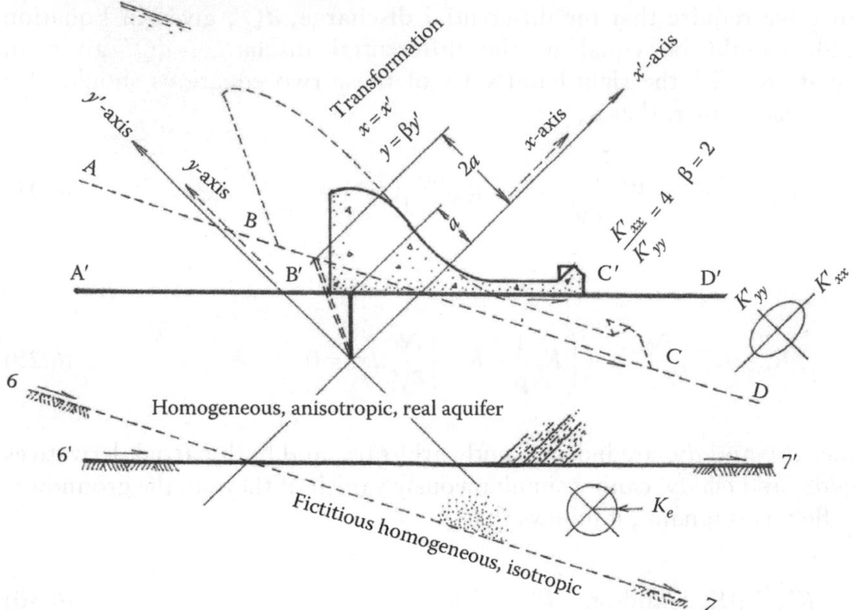

Figure 6.7 Mapping of real anisotropic flow region onto fictitious isotropic flow region.

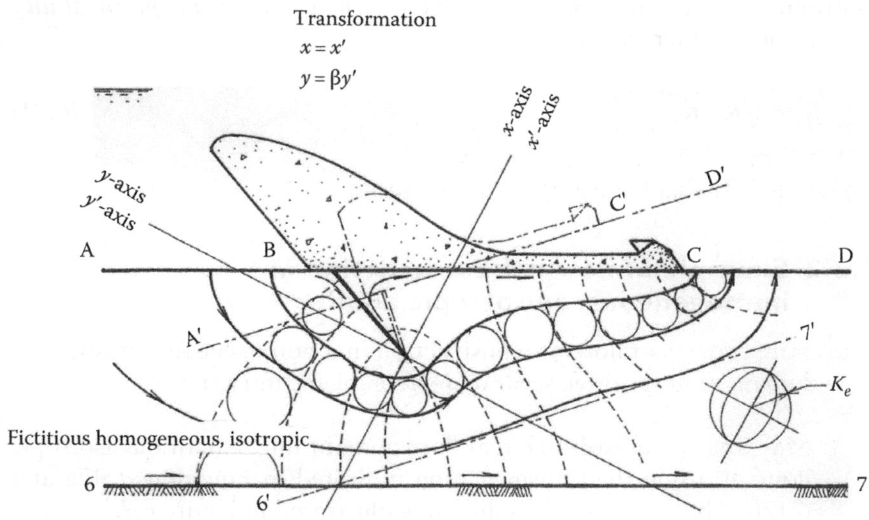

Figure 6.8 Flow net for fictitious isotropic flow region. *Notes:* Primed quantities refer to anisotropic aquifer. Unprimed quantities refer to fictitious isotropic flow domain. Coordinate axes are along principal directions of anisotropy of real aquifer.

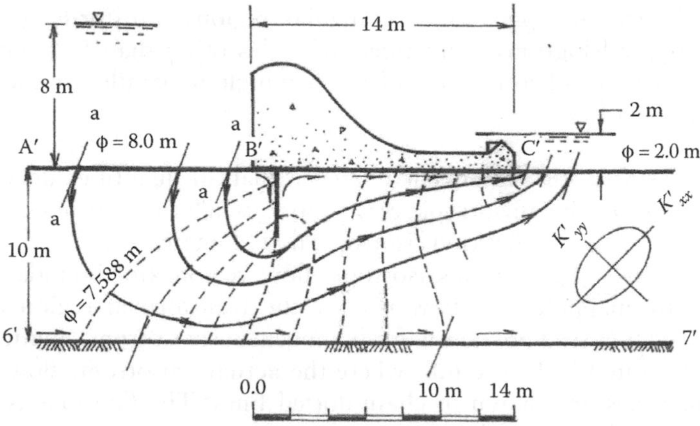

Figure 6.9 Flow net for anisotropic real aquifer.

3. Finally, finding the flow net (not necessarily orthogonal) for the actual anisotropic flow region using the inverse transformation and proceeding with the quantitative estimates, if so desired, for the seepage and the uplift forces on the structure. The illustration in Figure 6.9 represents the inverse transformation to obtain the corresponding flow net valid for the real anisotropic aquifer.

Step 1: The illustration shown in Figure 6.7 represents the actual anisotropic aquifer and its mapping onto a *fictitious isotropic* flow region. The original anisotropic flow region is shown in solid lines, while its transformed image is shown in dotted lines. For the implementation of the theory, we are only concerned with the mapping of the flow region and boundaries. The mapping of the hydraulic structures, or the free water surface, is not relevant. They are included in the transformation to add realism, besides providing a visual reference point to the otherwise abstract mathematical transformation. Using nonmathematical language, the transformation embodied in Equations 6.17 can be viewed as an expansion, or a contraction, of the actual flow region in the ordinate direction only. The coordinate axes are taken parallel to the principal directions of anisotropy of the real aquifer, as shown in Figure 6.7. For the purpose of illustration, the ratio between the principal values of coefficient of permeability is taken as $K'_{xx}/K'_{yy} = 4$, which results in the magnification factor $\beta = 2$. It is this magnification factor that has been used in transforming the real aquifer into a factitious isotropic flow region. Some general properties of this mapping are summarized as follows:

- Under this mapping, or transformation using Equations 6.17, the straight lines remain straight.
- Under this mapping, the parallel lines remain parallel.

- Under this mapping, the rectangular regions transform, generally, into parallelograms, with internal angles other than 90°. The only exception is when the side of the rectangle is parallel to one of the coordinate axes.

It must be stated for subsequent reference that all quantities denoted by prime (′) refer to the anisotropic real aquifer, while the unprimed denotation is reserved for the fictitious isotropic flow domain.

Step 2: Once the fictitious isotropic flow region and boundaries are obtained by mapping, the flow net for the transformed region can be obtained as discussed previously for homogeneously isotropic aquifer. This step is illustrated in Figure 6.8, where the actual anisotropic flow region and boundaries are shown in chain-dotted lines. The fictitious isotropic flow region is shown in solid lines.

Step 3: The (orthogonal) flow net obtained in Step 2 is transformed into the corresponding flow net (not necessarily orthogonal) for the real anisotropic aquifer using the inverse transformation. The result of this step is illustrated in Figure 6.9. It is important to emphasize that while the streamlines intersect the known equipotential AB at right angles in the fictitious isotropic flow region, they intersect the known equipotentials, such as A′B′, at an angle other than a right angle. This angle of intersection is shown typically by *aa* in Figure 6.9. It is however evident that all streamlines intersect the known equipotential A′B′ at the same angle, as typically shown by *aa*. The last assertion follows from the fact that parallel lines remain parallel during transformation.

In the case of an anisotropic aquifer, the quantitative analyses for seepage and uplift pressure follow the same steps as described for homogeneous, isotropic aquifer. It is left as an exercise for the reader to estimate the seepage and uplift pressure under the hydraulic structure shown in Figure 6.9.

6.2.3 Illustrative problem 6.1

For the structure shown in Figure 6.4, find the following: (a) the seepage through the aquifer, per width normal to the plane of paper, if the coefficient of permeability $K = 0.001$ cm/s; (b) the porewater pressure at point P; and (c) the uplift pressure due to porewater at the base of the hydraulic structure.

Part (*a*): The discharge in general is given by the following:

$$\text{Discharge} = (\text{Flow area}) \times (\text{Normal velocity}) \tag{IP6.1.1}$$

In this case, at a typical section of a flow channel, the flow area $= \Delta n \times 1$, and at the same section, the normal velocity is given by Darcy's law

$$q = -K\frac{d\phi}{ds} \cong K\frac{\phi_1 - \phi_2}{\Delta s\, n_\phi} \tag{IP6.1.2}$$

where n_ϕ denotes the number of equal drops in the piezometric head from the upstream boundary (AB) to the downstream boundary (CD). From Figure 6.4, this number is equal to 16. The discharge through a single flow channel is thus given as shown in the following:

$$\Delta Q = (\Delta n \times 1)K\frac{\phi_1 - \phi_2}{\Delta s\, n_\phi} \tag{IP6.1.3}$$

Since each cell of the flow net represents a curvilinear square, $\Delta n = \Delta s$. Thus, the aforementioned equation reduces to

$$\Delta Q = K\frac{\phi_1 - \phi_2}{n_\phi} \tag{IP6.1.4}$$

The total amount of flow per depth normal to the paper is, therefore, given by the following equation:

$$Q = n_\Psi \Delta Q = \frac{n_\Psi}{n_\phi}K(\phi_1 - \phi_2) \tag{IP6.1.5}$$

where n_Ψ denotes the total number of flow channels in Figure 6.4. In the aforementioned expression, the ratio n_Ψ/n_ϕ is called the shape factor (SF). It may be emphasized that neither n_Ψ nor n_ϕ necessarily represents a whole number. Furthermore, the SF for a well-drawn flow net does not depend on its further refinement. It is a property of the flow region and the boundary conditions. From Figure 6.4, it is estimated that $n_\Psi = 4.75$. Substituting the relevant values from the figure into Equation IP6.1.5 yields the required discharge, per width normal to the paper, as shown in the following:

$$\begin{aligned} Q &= \frac{4.75}{16} \times 0.001\,\text{cm/s} \times \left(\frac{1\,\text{m}}{100\,\text{cm}}\right) \times (8\,\text{m} - 2\,\text{m}) \\ &= 0.0000178\,\text{m}^2/\text{s} = 2.06\,\text{m}^2/\text{day} \end{aligned} \tag{IP6.1.6}$$

Part (*b*): The porewater pressure at any point P is given by the following expression:

$$\frac{p}{\gamma} = \phi - y \tag{IP6.1.7}$$

For point P, we find the following values:

$$\phi = 5.0 \text{ m}, y = -10 \text{ m} \tag{IP6.1.8}$$

Combining Equation IP6.1.8 with Equation IP6.1.7 yields the following expression for pressure at P:

$$\frac{p}{\gamma} = 5.0 - (-10) = 15 \text{ m (meter of water)} \tag{IP6.1.9}$$

Since p/γ represents the pressure inside the porewater, it acts equally in all directions, as shown in the following figure.

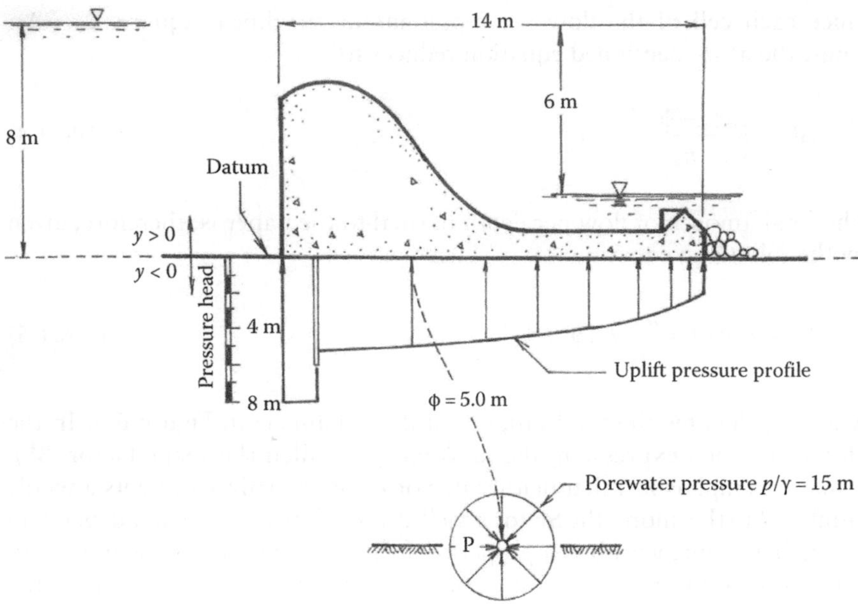

Part (c): Since the arbitrary datum for ϕ is taken at the same level as that of AB, the elevation head (y) is zero for all points located at the base of the hydraulic structure. Thus, at all points situated at the base of the hydraulic structure,

$$\frac{p}{\gamma} = \phi - y = \phi \tag{IP6.1.10}$$

Using flow net, the discrete points where the equipotential lines intersect the base of the hydraulic structures can be located. At these points, the uplift pressure can be found using Equation IP6.1.10. The uplift pressure profile is plotted at the base of the figure.

6.3 LAYERED HETEROGENEITY

A typical section of a stratified geological formation bearing groundwater is shown in Figure 6.10. For simplicity, each layer in this figure can be viewed as a homogeneous, isotropic medium with uniform thickness. However, the coefficient of isotropic permeability changes in general from layer to layer. Our objective here is to replace this medium of *layered heterogeneity* with an equivalent (in some sense) *homogeneous, anisotropic* medium. The term *equivalent* is deliberately left ambiguous, because its meaning depends on the objective of the analysis. In any case, the purpose of replacing the actual stratified formation with an equivalent medium tacitly assumes that such a replacement simplifies the mathematical formulation and thus reduces the analytical effort on the part of the analyst.

We assume a priori that the principal directions of the equivalent homogeneous, anisotropic aquifer align with the horizontal and vertical directions—the horizontal representing the direction of the bedding plane and the vertical representing the direction transverse to the bedding plane. We further assume that the equivalency implies that flows in the horizontal and vertical directions of the original formation are exactly the same as the corresponding flows in the equivalent homogeneous, anisotropic medium under the same hydraulic gradients. To accomplish this we proceed as follows:

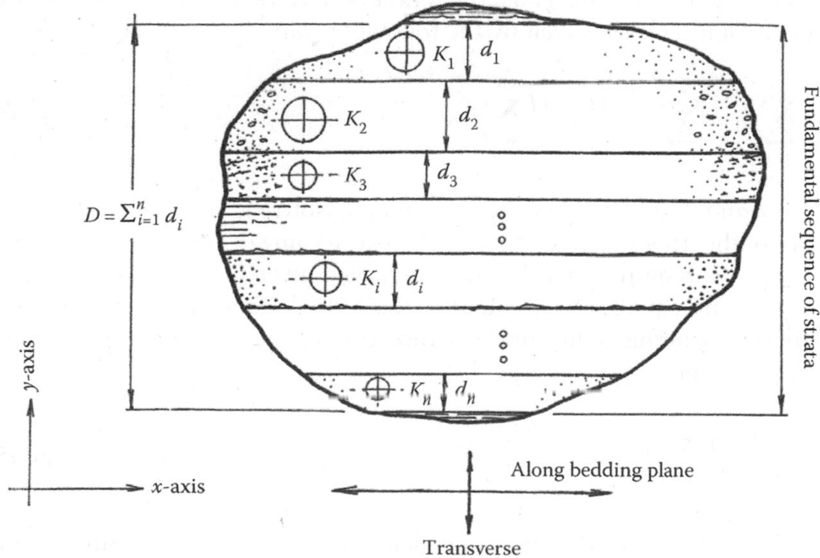

Figure 6.10 Layered heterogeneous geological formation.

Even for an anisotropic aquifer, Darcy's law in its original form is applicable along the principal direction (x-direction), because the specific discharge vector along the principal direction is normal to ϕ=constant surfaces. Thus, the discharge per width normal to the plane of paper, in an equivalent anisotropic aquifer, can be written, using Darcy's law, as

$$Q = A \times \left(-K'_{xx} \frac{d\phi}{dx} \right) \tag{6.32}$$

where

A denotes the flow area

K'_{xx} denotes the coefficient of permeability in the principal direction

$-\frac{d\phi}{dx}$ denotes the gradient of the piezometric surface in the principal direction

The aforementioned equation can also be written as

$$Q = (D \times 1) \times K'_{xx} \frac{H}{L} \tag{6.33}$$

where the quantities $(D \times 1)$ *and* H/L designate the flow area and the hydraulic gradient, respectively (Figure 6.11).

Now, with reference to Figure 6.11, it is noted that the hydraulic gradient is same for all layers. Thus, the total discharge per width normal to the plane of Figure 6.11 is given by the following sum:

$$\sum_{i=1}^{n} Q_i = \sum_{i=1}^{n} d_i K_i \frac{H}{L} = \frac{H}{L} \sum_{i=1}^{n} d_i K_i \tag{6.34}$$

where d_i and K_i denote the thickness and the isotropic coefficient of permeability of the ith layer, respectively. To ensure equivalency between the two aquifers, it is required that the two discharges given in Equations 6.33 and 6.34 must be equal. Thus, equating these discharges and expressing the coefficient of permeability in the x-direction of the equivalent anisotropic aquifer, we obtain

$$K'_{xx} = \frac{1}{D} \sum_{i=1}^{n} d_i K_i \tag{6.35}$$

Equation 6.35 shows that the equivalent coefficient of permeability in the x-direction is the thickness-weighted arithmetic average of individual coefficients of permeability, K_i.

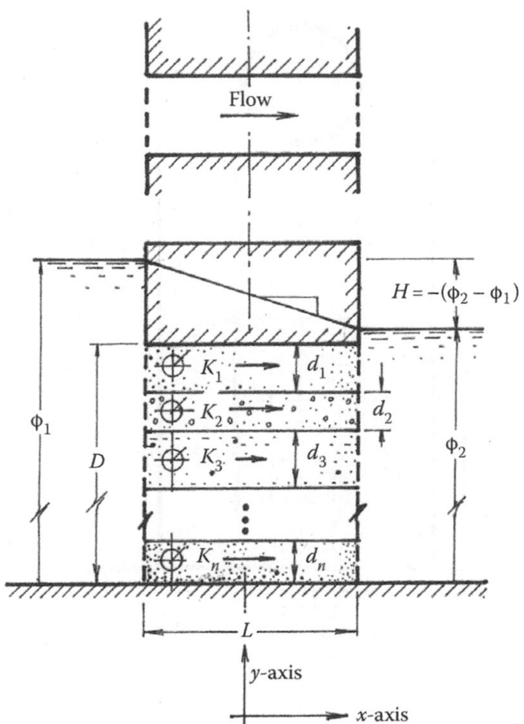

Figure 6.11 Definition sketch for flow in the *x*-direction.

Now, let us consider the flow in the vertical direction through the stratified medium, as shown in Figure 6.12. Again, Darcy's law in its original form is applicable because the *y*-axis is assumed to align with one of the principal directions of anisotropy of the equivalent aquifer. Thus, according to Darcy's law

$$q_y = -K'_{yy} \frac{d\phi}{dy} \tag{6.36}$$

where the symbols carry their usual meanings

q_y represents specific discharge (discharge per area) in the *y*-direction

K'_{yy} is the coefficient of permeability of an equivalent homogeneous, anisotropic aquifer

$-\dfrac{d\phi}{dy}$ represents the hydraulic gradient in the *y*-direction

Since the flow is steady and devoid of any point sinks or sources, the specific discharge in the *y*-direction through each layer must be the same; that is,

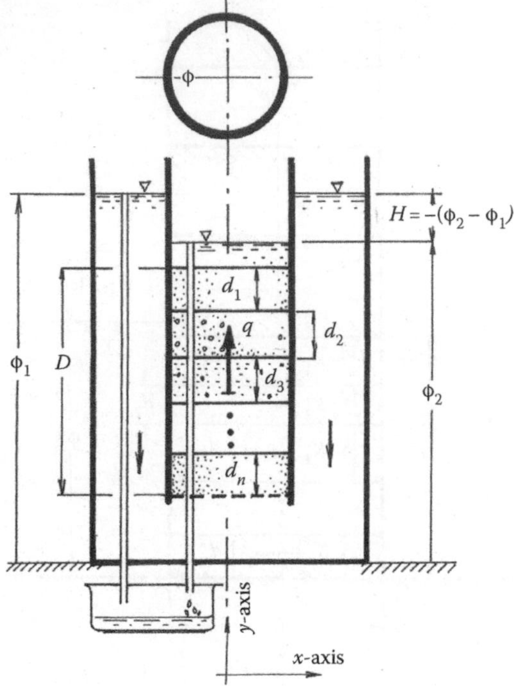

Figure 6.12 Definition sketch for flow in the *y*-direction.

$$q_{y1} = q_{y2} = q_{y3} = \cdots = q_{yn} = q_y \tag{6.37}$$

Equation 6.36 can be written as

$$q_y = K'_{yy} \frac{H}{D} \tag{6.38}$$

where H and D are defined in Figure 6.12. Since each layer is homogeneously isotropic with the coefficient of permeability, K_i, it is evident from Darcy's law that

$$\Delta H_i = \frac{q_{yi} d_i}{K_i} = q_y \frac{d_i}{K_i} \tag{6.39}$$

or

$$H = \sum_{i=1}^{n} \Delta H_i = q_y \sum_{i=1}^{n} d_i \left(\frac{1}{K_i} \right) \tag{6.40}$$

Combining Equation 6.40 with Equation 6.38 yields the required expression for the coefficient of permeability in the y-direction for the equivalent homogeneously anisotropic aquifer, as shown in the following:

$$\frac{1}{K'_{yy}} = \frac{1}{D}\sum_{i=1}^{n} d_i \left(\frac{1}{K_i}\right) \tag{6.41}$$

The last expression shows that the equivalent coefficient of permeability in the y-direction represents the thickness-weighted harmonic average of individual coefficients of permeability. It is known that the arithmetic average is always greater than the harmonic average. In other words, $K'_{xx} > K'_{yy}$. For a demonstration of the truth of this assertion, see Harr (1991, p. 28) or Terzaghi (1943, p. 244). Further discussion on this subject can be found in Charbeneau (2000). This result shows that the direction parallel to the bedding planes represents the major principal axis, while the direction normal to the bedding planes represents the minor principal axis of the equivalent homogeneous, anisotropic aquifer.

These results are graphically summarized in Figure 6.13, which represents side-by-side the *layered isotropic heterogeneity* of the geological

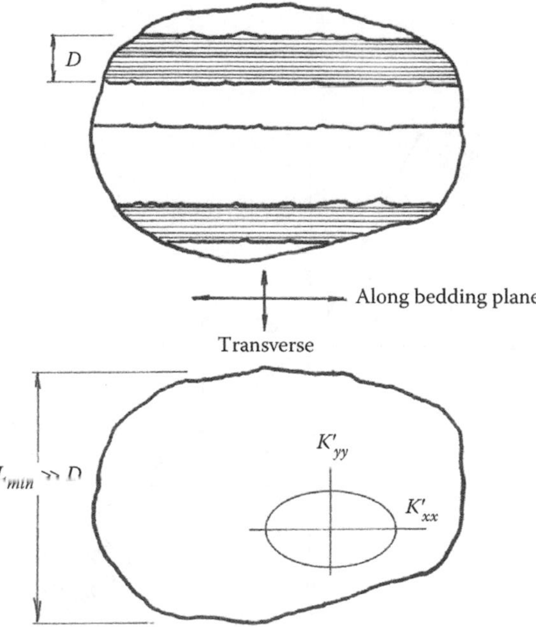

Figure 6.13 Graphical representation of equivalency.

formation and its equivalent *homogeneously anisotropic* formation. In the lower figure, the characteristic length (L_{min}) is assumed much larger than the total depth of the fundamental sequence of strata (Figure 6.10); otherwise, the whole idea of equivalency becomes meaningless.

6.3.1 Refraction of streamline

When a streamline enters a new homogeneous, isotropic aquifer, it experiences, in general, an abrupt change in its direction due to refraction. This refraction of streamline at the interface of two different isotropic strata is similar (though not identical) to the refraction of a ray of light in optics. To understand the physics of this phenomenon, Figure 6.14 has been prepared. This figure shows the interface between two homogeneous, isotropic strata of differing coefficients of permeability, K_1 and K_2. The variables n and s designate the normal and tangential coordinates, respectively, at any point on the interface. Since there are no point sources or sinks within the flow field, and the flow is in a steady-state condition, it follows that the component of the specific discharge vector, normal to the interface, q_n, must be continuous at all points on the interface. Thus (see Figure 6.14),

$$q_{n1} = q_{n2} \tag{6.42a}$$

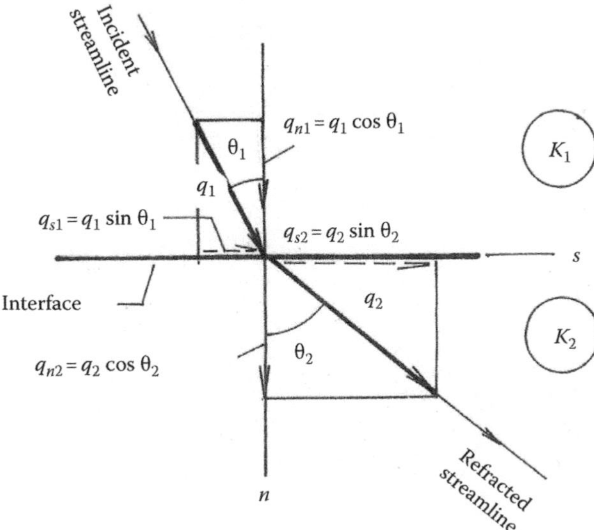

Figure 6.14 Refraction of streamline at the interface of different isotropic aquifers.

$$q_1 \cos\theta_1 = q_2 \cos\theta_2 \tag{6.42b}$$

$$\frac{q_1}{q_2} = \frac{\cos\theta_2}{\cos\theta_1} \tag{6.42c}$$

where the subscript 1 and 2 refer to the two strata with coefficients of permeability K_1 and K_2, respectively. It is also concluded that the pressure field is continuous at every point of the flow field (including the point lying on the interface), for, otherwise, it will cause an infinite pressure gradient that cannot be sustained by the aquifer. This leads to the conclusion that the piezometric head ϕ_1, in aquifer 1, and piezometric head ϕ_2, in aquifer 2, should be equal at any point located on the interface between the two strata. In other words, the piezometric head, $\phi(x, y) = p/\gamma + y$, is continuous across the interface, because p/γ as well as y are continuous across the interface. Also, the tangential components of the specific discharge vectors, q_1 and q_2, along the two adjacent sides of the interface are given by the following:

$$q_1 \sin\theta_1 = -K_1 \frac{\partial \phi_1}{\partial s} = -K_1 \frac{\partial \phi}{\partial s} \tag{6.43a}$$

$$q_2 \sin\theta_2 = -K_2 \frac{\partial \phi_2}{\partial s} = -K_2 \frac{\partial \phi}{\partial s} \tag{6.43b}$$

where, in the second equality, we have used the fact that $\partial\phi_1/\partial s = \partial\phi_2/\partial s =$ (say, $\partial\phi/\partial s$) due to the continuity of the piezometric head across the interface. Thus, division of Equation 6.43a by Equation 6.43b yields the following:

$$\frac{q_1}{q_2} = \frac{K_1}{K_2} \frac{\sin\theta_2}{\sin\theta_1} \tag{6.44}$$

Eliminating q_1/q_2 from Equations 6.44 and 6.42c, we obtain

$$\frac{K_1}{K_2} = \frac{\tan\theta_1}{\tan\theta_2} \tag{6.45}$$

This shows that during refraction, an incident streamline undergoes an abrupt change in direction from θ_1 to θ_2 (Figure 6.14). The change is characterized by the fact that the ratio of tangents of angles the streamline makes with the normal to interface is exactly the same as the permeability ratio. With the exception of $\theta_1 = 0$, larger the difference between

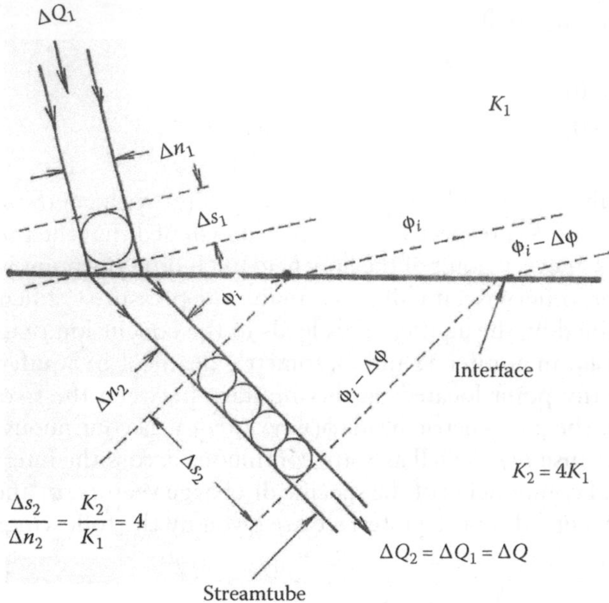

Figure 6.15 Consequence of refraction of streamlines.

the coefficients of permeability, larger is the angular deviation of the streamline after refraction. Further consequences of Equation 6.45 are illustrated graphically in Figure 6.15.

6.4 CONCLUDING REMARKS

The flow net in fact represents the graphical solution of the boundary-value problem associated with the Laplace equation $\nabla^2\phi = 0$ subject to the known boundary conditions. *It may be emphasized here that for a given flow region and boundaries, the flow net is unique and does not depend on the actual numerical specification of the boundary conditions.* In other words, the flow net configuration remains the same, as long as the flow region and the type of the boundaries (with regard to constancy of ϕ and Ψ) remain the same. This aspect of the flow net analysis is particularly appealing in the early studies of a groundwater project. It is the relative ease with which the flow net can be constructed and analyzed that makes its use so attractive. However, all aquifers are heterogeneous and anisotropic to a varying degree of complexity. The flow nets presented in the foregoing sections pertain to rather simplified models of aquifer. Despite these inherent simplifications, the flow net analyses provide an additional degree of confidence to the judgment of a groundwater hydrologist or a geotechnical engineer.

6.5 EXERCISES

6.1 Find the seepage per width normal to the plane of figure of the structure shown in Figure 6.9 if the coefficients of permeability are $K'_{xx} = 0.0004$ m/s and $K'_{yy} = 0.0001$ m/s.

6.2 Estimate the uplift force due to porewater pressure on the hydraulic structure shown in Figure 6.9.

6.3 Find the relationship among α, γ, β so that the inclined sheet pile becomes vertical after the anisotropic, homogeneous aquifer $\beta = \sqrt{K'_{xx}/K'_{yy}}$ is transformed into a fictitious isotropic, homogeneous flow domain. Assume $0 < \alpha < \gamma < 1/2\,\pi$.

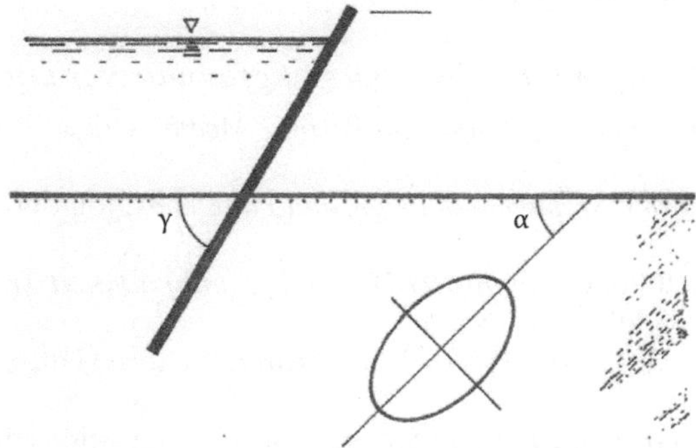

6.4 For the situation shown in the following figure (flow net partly drawn), find the following: (1) the seepage through the aquifer per width normal to the plane of paper and (2) the pressure profile on both sides of the sheet pile. Assume the aquifer is homogeneous, isotropic and the coefficient of permeability $K = 0.005$ cm/s.

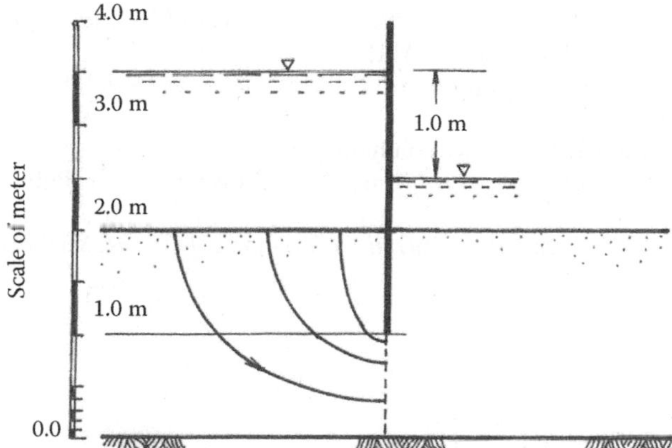

6.5 During refraction, an incident streamline undergoes an abrupt change in direction as shown in Figures 6.14 and 6.15. With regard to variables defined in these figures, prove the following:

(a) $\dfrac{\Delta s_2}{\Delta s_1} = \dfrac{\sin\theta_2}{\sin\theta_1}$

(b) $\dfrac{\Delta n_2}{\Delta n_1} = \dfrac{\cos\theta_2}{\cos\theta_1}$

SUGGESTED READINGS

Related to flow nets and the mechanics of groundwater flow in general:

Bear, J. 1988. *Dynamics of Fluids in Porous Media*, Dover Publications, Inc., New York.

Cedergren, H. R. 1989. *Seepage, Drainage, and Flow Nets*, John Wiley & Sons, Inc., New York.

Charbeneau, R. J. 2000. *Groundwater Hydraulics and Pollutant Transport*, Prentice Hall, Upper Saddle River, NJ.

Freeze, R. A. and J. A. Cherry. 1979. *Groundwater*, Prentice Hall, Englewood Cliffs, NJ.

Harr, M. E. 1991. *Groundwater and Seepage*, Dover Publications, Inc., New York.

Hubbert, K. M. 1940. The theory of ground-water motion, *Journal of Geology*, 48, 785–944.

Strack, O. D. L. 1989. *Groundwater Mechanics*, Prentice Hall, Inc., Englewood Cliffs, NJ.

Terzaghi, K. 1943. *Theoretical Soil Mechanics*, John Wiley & Sons, Inc., New York.

Todd, D. K. 1959. *Ground Water Hydrology*, John Wiley & Sons, Inc., New York.

Mathematics applied to the Laplace equation:

Churchill, R. V. and J. W. Brown. 1990. *Complex Variables and Applications*, 5th edn., McGraw-Hill Inc., New York.

Kellogg, O. D. 1953. *Foundations of Potential Theory*, Dover Publications Inc., New York.

Rothe, R., Ollendorff, F., and K. Pohlhausen (Eds.). 1961. *Theory of Functions as Applied to Engineering Problems*, Dover edition, Dover Publications Inc., New York.

Sneddon, I. N. 1957. *Elements of Partial Differential Equations*, McGraw-Hill Book Company, Inc., New York.

Chapter 7

Determination of aquifer characteristics

In this chapter, we present from a pedagogical point of view the theory behind in situ determination of two aquifer characteristics: the transmissivity and the storativity of the aquifer. We shall first define these terms for confined aquifers and later extend the definition to unconfined aquifers. For a confined aquifer, the transmissivity is defined as $T=KH$, where T denotes the transmissivity, K the coefficient of permeability of a homogeneous, isotropic aquifer, and H the uniform depth of a horizontal confined aquifer. The storativity, S, of a horizontal confined aquifer represents the amount of water released or stored per aquifer area per unit change in the piezometric head. Storativity, S, is a dimensionless quantity, and in the case of confined aquifers, its numerical value is rather low—generally less than 0.001. These characteristics are further shown graphically in Figure 7.1, which illustrates the physical meaning of transmissivity in a confined aquifer and that of storativity in an unconfined aquifer. The transmissivity can be interpreted as the discharge per width through the entire aquifer depth under a unit hydraulic gradient (Figure 7.1a). The storativity on the other hand depends on two separate physical properties: the compressibility of water and the squeezing (consolidating) property of the aquifer. The amount of water released due to the decline of pressure depends on the compressibility of water as well as the consolidation of the aquifer. Thus, the storativity of an aquifer, in principle, can be calculated in terms of compressibility of water and the consolidation characteristics of the aquifer.

In case of unconfined aquifer, the transmissivity, T, is in general a variable quantity and can only be defined as an average property. On the other hand, the storativity of an unconfined aquifer can be closely approximated simply by its porosity (V_v/V). This fact is illustrated in Figure 7.1b.

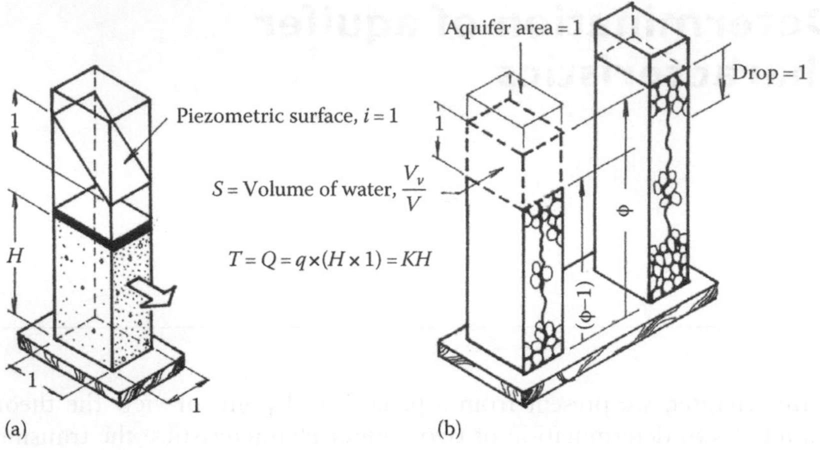

Figure 7.1 (a) Transmissivity in a confined aquifer and (b) storativity in an unconfined aquifer.

7.1 DETERMINATION OF TRANSMISSIVITY OR COEFFICIENT OF PERMEABILITY

7.1.1 Thiem equation: Confined aquifer

Before we discuss the determination of transmissivity, it is relevant to argue about the notion of a steady-state (equilibrium) groundwater flow field in relation to the Thiem equation. The Thiem equation for confined aquifers was derived in Chapter 4 and is reproduced here for quick reference:

$$\phi(r) = \frac{Q}{2\pi KH} \ln\frac{r}{r_0} + \phi_0 \tag{4.33}$$

In the aforementioned equation, the symbols carry their usual meanings. The symbol ϕ_0 denotes the constant (not changing with time) piezometric head at radial distance r_0 from the pumping well. There are a number of points that should be emphasized with regard to this equation. The foremost among these is the fact that the derivation of the Thiem equation is based on the assumption of a steady-state flow condition. In a horizontal, extensive confined aquifer, it takes an infinite time to develop a steady-state condition. However, in practice, after a sufficient lapse of time since the beginning of pumping, the potential field around the well appears to achieve a quasi-limiting condition in which the change in the piezometric surface becomes imperceptible. Under this condition, the previous equation can be written in the following forms:

$$\phi(r_1) = \frac{Q}{2\pi KH} \ln \frac{r_1}{r_2} + \phi(r_2) \tag{7.1a}$$

$$T = KH = \frac{Q}{2\pi\left[\phi(r_1) - \phi(r_2)\right]} \ln \frac{r_1}{r_2} \tag{7.1b}$$

Equation 7.1b can be used for the determination of transmissivity of a confined aquifer. Although Equation 7.1b is strictly valid for a steady-state condition, its practical utility for the determination of transmissivity, despite a quasi-steady-state condition, is recognized in the literature (see, for instance, Todd, 1959, p. 83). And its use in aquifer testing is later justified in Section 7.6. In Equation 7.1b, the variables r_1 and r_2 denote the radial distances from the pumping well to two observation wells, with piezometric elevations $\phi(r_1)$ and $\phi(r_2)$, respectively.

7.1.2 Dupuit equation: Unconfined aquifer

Similar to the Thiem equation, the steady-state potential function for an unconfined, homogeneous, isotropic aquifer was derived in Chapter 4, based on the Dupuit–Forchheimer assumptions. The Dupuit equation can be written in the following forms:

$$\phi_1^2 - \phi_2^2 = \frac{Q}{\pi K} \ln \frac{r_1}{r_2} \tag{7.2a}$$

$$K = \frac{Q}{\pi\left[\phi_1^2 - \phi_2^2\right]} \ln \frac{r_1}{r_2} \tag{7.2b}$$

Equation 7.2b can be used to determine the coefficient of permeability K of the unconfined, homogeneous, isotropic aquifer. In the previous equations, the symbols ϕ_1 and ϕ_2 denote the phreatic surface elevations at observation wells located at radial distances r_1 and r_2, respectively.

As a practical advice, it is recommended that the locations of observation wells should neither be very close nor too far from the pumping well so that the changes in the elevation of the piezometric (or phreatic) surface could be determined with reasonable precision. Finally, the steady-state equations discussed hitherto for confined and unconfined aquifers do not provide any information about the storativity of aquifers. This limitation is, however, overcome by the use of transient analysis, which constitutes the subject matter of the proceeding sections. Despite these limitations, the steady-state equations are regarded to yield better estimations of aquifer transmissivity or coefficient of permeability than transient analyses (Todd, 1959; Fetter, 2001).

7.1.3 Illustrative problem 7.1

A fully penetrating well in a homogeneous, isotropic, horizontal confined aquifer of depth 15 m. is pumped at a constant rate of $Q = 1000$ m^3/day. After sufficient time, the piezometric surface appears to be in a quasi-limiting state. At that time, the piezometric levels in two observation wells, located at radial distances 10 m and 20 m, are 20 m and 22 m, respectively (see the following figure). Find (a) the transmissivity and (b) the coefficient of permeability of the aquifer.

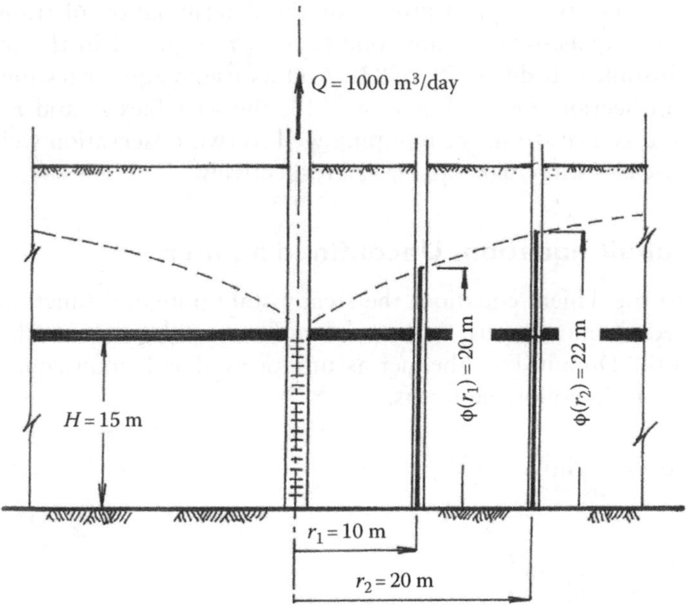

Solution: Since the problem pertains to a confined aquifer, Equation 7.1b is applicable. Thus, substituting the appropriate data in Equation 7.1b yields the following:

$$T = KH = \frac{Q}{2\pi\left[\phi(r_2) - \phi(r_1)\right]} \ln \frac{r_2}{r_1}$$

$$= \frac{1000\,\text{m}^3/\text{day}}{2\pi(22\,\text{m} - 20\,\text{m})} \ln \frac{20}{10} = 55.155\,\text{m}^2/\text{day}$$

The coefficient of permeability K can be obtained as shown in the following:

$$K = \frac{T}{H} = \frac{55.155\,\text{m}^2/\text{day}}{15\,\text{m}} = 3.68\,\text{m/day}$$

7.2 THEIS EQUATION: TRANSIENT RADIAL FLOW TO A WELL IN A CONFINED AQUIFER

We shall consider here the unsteady (transient) radial flow to a fully penetrating well in a horizontal confined aquifer. The discussion is based on the solution given by Theis (1935). Theis obtained the solution for the transient radial flow toward a well based on an analogy between the groundwater flow and the flow of heat by conduction. The Theis solution is based on the following assumptions:

1. The aquifer is homogeneous, isotropic, and confined by two horizontal impermeable layers of infinite areal extent, but at a finite distance, H, apart.
2. The original piezometric surface (or potentiometric surface) prior to pumping is also horizontal at some finite height, ϕ_0, from an arbitrary datum.
3. The well is pumped at a constant rate, Q, and its diameter is infinitely small.
4. The well axis is vertical, or the flow is planar radial.
5. The groundwater is released instantaneously by the aquifer, and the celerity of the pressure wave (disturbance) is infinite.

Under the aforementioned assumptions, the solution of transient flow to the well must satisfy the following continuity equation:

$$\frac{\partial^2\phi(r,t)}{\partial r^2}+\frac{1}{r}\frac{\partial\phi(r,t)}{\partial r}=\frac{S}{T}\frac{\partial\phi(r,t)}{\partial t} \tag{7.3}$$

In the aforementioned equation, the height of piezometric (or potentiometric) surface, $\phi(r, t)$, above an arbitrary datum is a function of two independent variables: r and t. The variable r denotes the radial distance as before, and the variable t denotes the lapsed time since the initiation of pumping, at a constant rate, Q. Since, for steady-state cases, the piezometric height ϕ is not a function of time, the preceding equation reduces to Laplace equation in polar coordinates (r, θ), for steady, planar, radial flows. Figure 7.2 represents the definition of flow field as well as the pertinent variables associated with the transient analysis of flow toward a fully penetrating well in a homogeneous, isotropic, horizontal confined aquifer. The variable,

$$s(r,t)=\phi_0-\phi(r,t)$$

as shown in Figure 7.2, is called the *drawdown*. At any given time, the three-dimensional surface traced by $s(r, t)$ with radial symmetry is called

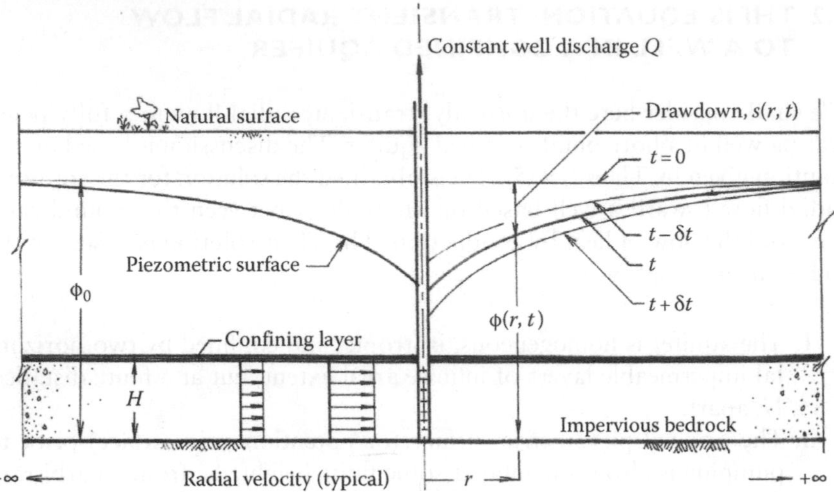

Figure 7.2 Definition sketch for the Theis equation.

the *cone of depression*. In transient flows, this cone of depression of the piezometric (or potentiometric) surface deepens with the passage of time.

In order to solve the partial differential Equation 7.3, we need initial and boundary conditions. These are provided as follows:

$$\phi(r,0) = \phi_0 \quad \text{for all } r\text{'s} \tag{7.4a}$$

$$\phi(r,t) \to \phi_0 \quad \text{as } r \to \infty \text{ for all } t\text{'s} \tag{7.4b}$$

$$\lim_{r\to 0}\left[r\frac{\partial \phi}{\partial r}\right] = \frac{Q}{2\pi T} \tag{7.4c}$$

Equation 7.4a represents the initial condition, and Equation 7.4b describes the behavior of the potential function ϕ at the far field. This description also implies that the cone of depression only deepens with time, and its rim (or periphery) theoretically remains at infinity for all times. The boundary condition (7.4c) maintains the continuity of flow at the well, in spite of the fact that the well radius tends to zero.

The mathematical solution to the boundary-value problem formulated hitherto has been obtained by Theis (1935). And it can be stated as follows:

$$s(r,t) = \phi_0 - \phi(r,t) = \frac{Q}{4\pi T}\int_u^\infty \frac{1}{x}e^{-x}dx \tag{7.5a}$$

$$u \equiv \frac{S}{4T}\left[\frac{r^2}{t}\right] \tag{7.5b}$$

The variable x in the exponential integral is the dummy variable of integration. The value of this integral depends on the lower limit of integration, u. Here, u is a dimensionless parameter, and for a given aquifer, it depends only on two variables, r and t, because T and S are known or fixed parameters. In the context of groundwater mechanics, the integral in Equation 7.5a is called the *Theis well function* and is denoted by $W(u)$. Thus,

$$s(r,t) = \phi_0 - \phi(r,t) = \frac{Q}{4\pi T} W(u) \tag{7.6}$$

The Theis well function can be expressed by the following infinite series:

$$W(u) = -0.5772 - \ln(u) + u - \frac{u^2}{2\times 2!} + \frac{u^3}{3\times 3!} - \frac{u^4}{4\times 4!} + \cdots \tag{7.7}$$

It is readily seen from the preceding equation that W is a dimensionless function of the dimensionless argument u. This well function, W, has been tabulated in terms of its argument u by Wenzel (1942). Although tables depicting numerical values of W in terms of u appear in other standard texts on groundwater hydrology (see, for instance, Todd, 1959; Freeze and Cherry, 1979; Carbeneau, 2000; Fetter, 2001), Appendix C includes this table, after Wenzel, for quick reference. It should also be mentioned that Carbeneau (2000) has included the spreadsheet module based on Basic computer programming language in his book.

The well function W is shown graphically in Figure 7.3. The curve in this figure is called the *reverse type curve*. Traditionally, abscissa in this figure represents $1/u$, instead of u. *The most important fact, however, about this curve is that it is a* ***universal graph*** *in the sense that any and all transient flows in an ideal Theis aquifer satisfy this curve irrespective of the actual values of variables, r and t, and the values of parameters, Q, T, and S*. It is this fact that is rightfully exploited later in the solution of the inverse problem by the *Theis method* for the determination of aquifer characteristics, T and S. In the present context, the inverse problem is defined as follows: When the drawdown at a radial distance at a given time since the beginning of pumping is known, the problem to find the aquifer characteristics, T and S, is called the inverse problem.

Equations 7.6, 7.7, and 7.5b are called *nonequilibrium equations* or *Theis equations* for the analysis of transient planar radial flows toward a fully penetrating well in an extensive, confined aquifer. In a given case, the successive development of drawdown of piezometric surface with time can be obtained from these equations.

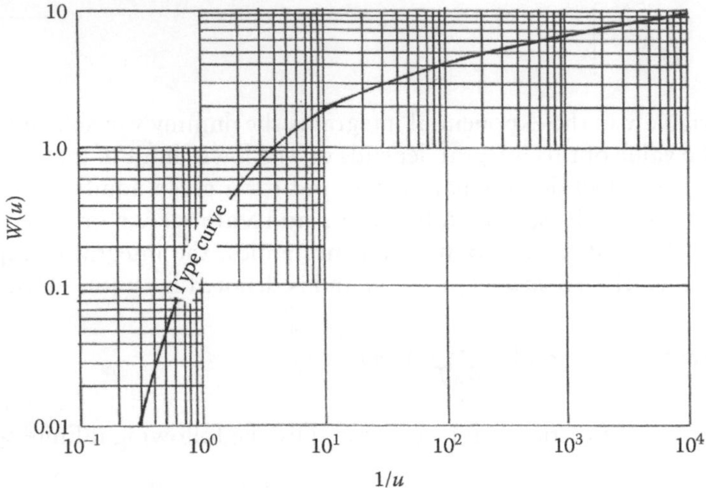

Figure 7.3 Reverse type curve.

7.2.1 Illustrative problem 7.2

A fully penetrating well in a confined aquifer is pumped at a constant rate, 2000 m³/day. Assuming a homogeneous, isotropic, horizontal confined aquifer with $T=200$ m²/day and $S=0.0005$, find the drawdowns at distances 6 and 12 m from the well, after 2 days of pumping.

Solution: Find parameter u and $W(u)$ at $r=6$ m.

$$u = \frac{S}{4T}\left[\frac{r^2}{t}\right] = \frac{0.0005}{4 \times 200\ \mathrm{m^2/day}}\frac{(6\ \mathrm{m})^2}{2\ \mathrm{days}} = 0.00001125$$

$W=10.82$, *from first two term approximation of series* (7.7)

Find parameter u and $W(u)$ at $r=12$ m.

$$u = \frac{S}{4T}\left[\frac{r^2}{t}\right] = \frac{0.0005}{4 \times 200\ \mathrm{m^2/day}}\frac{(12\ \mathrm{m})^2}{2\ \mathrm{days}} = 0.000045$$

$W=9.43$, *from first two term approximation of series* (7.7)

The drawdowns can be obtained from the equation $s(r, t)=(Q/4\pi T)W$ by substituting the appropriate values. Thus,

$$s(6\ \mathrm{m}, 2\ \mathrm{days}) = \frac{2000\ \mathrm{m^3/day}}{4\pi \times 200\ \mathrm{m^2/day}} \times 10.82 = 8.61\ \mathrm{m}$$

$$s(12\,\text{m}, 2\,\text{days}) = \frac{2000\ \text{m}^3/\text{day}}{4\pi \times 200\ \text{m}^2/\text{day}} \times 9.43 = 7.50\ \text{m}$$

Remarks: This illustrative problem shows that the drawdown at any time at any radial distance due to constant pumping can be obtained from the Theis equation, provided the aquifer constants, T and S, are known a priori. More important from a practical perspective is the inverse problem.

7.3 THEIS METHOD

Theis proposed a graphical method for the determination of aquifer constants, when the observed values of drawdowns at various time points are known either for a single or multiple (nonpumping) observation wells. The method is based on the Theis equations for transient response of the idealized Theis aquifer to the constant rate of pumping. The theory and procedure of the Theis method are given in the following.

Taking the logarithm of Equations 7.6 and 7.5b yields the following results:

$$\log\left[s(r,t)\right] = \log\left[\frac{Q}{4\pi T}\right] + \log\left[W(u)\right] \tag{7.8}$$

$$\log\left[\frac{t}{r^2}\right] = \log\left[\frac{S}{4T}\right] + \log\left[\frac{1}{u}\right] \tag{7.9}$$

If the discharge well is pumped at a constant rate, the quantity Q in Equation 7.8 remains a constant. Thus, for a fully penetrating well in a confined aquifer with constant Q, the term $\log[Q/4\pi T]$ also remains constant, because T is constant for a homogeneous confined aquifer. This further implies that the two variables $\log[s(r, t)]$ and $\log[W(u)]$ are essentially the same, with the exception that they differ by a constant. A similar reasoning also holds with regard to variables $\log[t/r^2]$ and $\log[1/u]$ in Equation 7.9, because S and T are constant for a given homogeneous confined aquifer.

The previous discussion has some further implications. For instance, the plot of W versus $1/u$ is similar in shape to the plot of $s(r, t)$ versus t/r^2, when drawn on log-log graph papers of same scale and size. The graph of W versus $1/u$ is a unique curve, which is theoretically obtained and does not rely on the field observation for its determination.

The graph between $s(r, t)$ and t/r^2 is called the *field data plot* (or, simply, the *data plot*), and it does depend on the actual field observations. If drawn to the same scale as the reverse type curve (Figure 7.3), the data plot should look identical to the reverse type curve. These two curves can further be

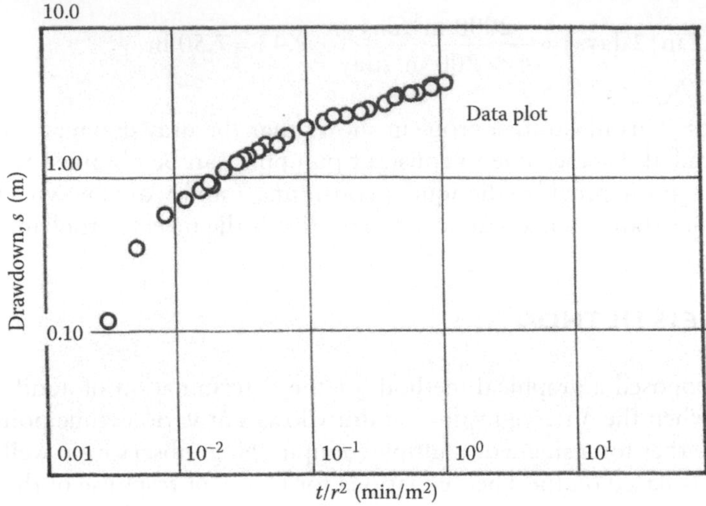

Figure 7.4 Field data plot.

brought to a near-congruence by sliding one curve relative to the other while keeping the respective abscissa and ordinate axes of the two graphs parallel.

A step-by-step procedure for the determination of aquifer characteristics, using the Theis method, is outlined as follows:

1. We begin with the construction of the reverse type curve, once for all, on a log-log graph paper as shown in Figure 7.3.
2. For each aquifer testing, the field observations are recorded to construct the data plot. For the construction of the data plot, the drawdowns $s(r, t)$ are plotted against the variable t/r^2 on a log-log graph paper with same scale and size as that of the type curve (Figure 7.4). It must be emphasized that the drawdown $s(r, t)$ is plotted as the ordinate, and the variable t/r^2 as the abscissa. This ensures that the variable $\log[s]$ corresponds to $\log[W]$, while the variable $\log[t/r^2]$ corresponds to $\log[1/u]$.
3. The data plot is slid relative to the type curve to achieve visually the best fit between the data and the type curve, while maintaining the respective axes parallel (see Figure 7.5). For this step, a light table may be handy.
4. Once the best fit between the data and the type curve is obtained, any convenient point (not necessarily on the curve) may be selected as the match point (see point P, in Figure 7.5). The ordinate and the abscissa of the match point are recorded on both graphs, as shown in the figure.

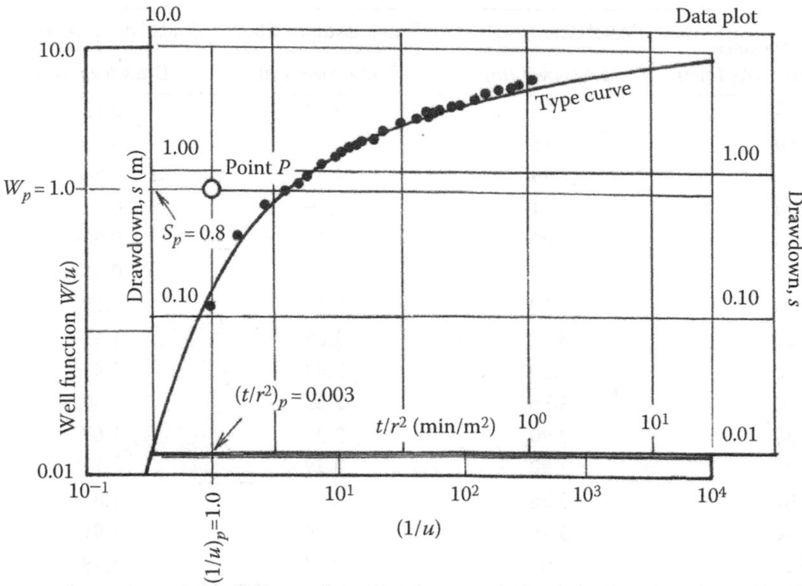

Figure 7.5 Superimposing the data plot on the type curve to achieve best fit.

5. Finally, the aquifer constants are obtained from the following equations:

$$T = \frac{QW_p}{4\pi s_p}$$

$$S = 4T\left[\frac{t}{r^2}\right]_p \div \left[\frac{1}{u}\right]_p$$

In these equations, subscript p refers to the respective coordinates of the match point P (Figure 7.5).

The details of the aforementioned procedure are further described in the following illustrative problem.

7.3.1 Illustrative problem 7.3

During an aquifer testing, the time–drawdown data were obtained as shown in the following table. Using the Theis method, determine the aquifer constants, T and S, if the pumping rate, $Q = 500$ m³/day, were kept constant during the aquifer testing.

Time since pumping (min)	Radial distance 10 m	Radial distance 20 m	Radial distance 40 m
	Drawdown (m)	Drawdown (m)	Drawdown (m)
5	1.75	0.72	0.13
10	2.15	1.19	0.35
15	2.39	1.43	0.56
20	2.71	1.59	0.72
25	2.80	1.83	0.80
30	3.02	1.99	0.95
40	3.18	2.23	1.19
50	3.42	2.39	1.35
60	3.50	2.43	1.43
70	3.58	2.55	1.59
80	3.90	2.71	1.67
90	3.98	2.79	1.71
100	3.98	2.82	1.83
110	3.98	2.86	1.91
120	4.02	2.94	1.99

Solution: We should use consistent units—actual units are arbitrary but consistency is not! We decide to use minutes for measuring time. Thus,

$$Q = 500\,\text{m}^3/\text{day} = 0.3472\,\text{m}^3/\text{min}$$

The reverse type curve for the Theis method is constructed once for all, as shown in Figure 7.3. Using the data shown in the table, the drawdown $s(r, t)$ is plotted against the variable (t/r^2) on the log-log paper with the same scale as that used for the type curve. This plot is shown in Figure 7.4. The next task is to obtain the best fit between the type curve and the data curve. For this purpose, we take (arbitrarily) the match point on the type curve with coordinates (1, 1). After achieving the best fit visually, we record the following correspondence between the variables (see Figure 7.5):

$W_p = 1$ corresponds to $s_p = 0.8$ m

$(1/u)_p = 1$ corresponds to $(t/r^2)_p = 0.003$ min/m^2

Substituting the appropriate values, we obtain

$$T = \frac{QW_p}{4\pi s_p} = \frac{0.3472\ \text{m}^3/\text{min}}{4\pi(0.8\ \text{m})} = 0.0388\ \text{m}^2/\text{min}$$

and

$$S = 4T\left(\frac{t}{r^2}\right)_p \div \left(\frac{1}{u}\right)_p$$

$$= 4 \times \left(0.0388 \text{ m}^2/\text{min}\right) \times \left(0.003 \text{ min/m}^2\right) = 0.00047$$

7.3.2 Optimization problem

The Theis method for the determination of aquifer characteristics is essentially a graphical method. Although it is extremely useful in understanding the basic theory behind the procedure, it is not very convenient in the contemporary computational environment of engineering offices. It is therefore desirable to cast this graphic approach into numerical computational procedures, using ubiquitous availability of personal computers. The Theis method can be easily formulated as a nonlinear optimization problem. The optimization problem can be stated as follows:

Find a solution vector $\begin{Bmatrix} T \\ S \end{Bmatrix}$, which meets the following explicit constraints

$$T_{min} \leq T \leq T_{max}$$

$$S_{min} \leq S \leq S_{max}$$

and minimizes the following error function (or objective function)

$$e = \frac{1}{n}\sum_{i=1}^{n}\left(s_i^c - s_i^o\right)^2$$

where

Subscript i refers to the observed data point

Superscripts c and o refer to the computed and observed values of drawdown, respectively

Subscripts min and max refer to the minimum and the maximum constraints on the values of T and S

The aforementioned optimization problem can be solved using standard methods. For instance, the Complex method of Box can be easily used (Box, 1965; Haque, 1985). For a given solution vector $\begin{Bmatrix} T \\ S \end{Bmatrix}$ and Q, the computed value of drawdown can be determined using the Theis Equations (7.6), (7.7), and (7.5b). This requires the numerical determination of the well function. The interested reader may consult Appendix D, Carbeneau (2000), or Strack (1989, pp. 196–200) for the computer implementation of the approximate formulas for the well function.

7.4 JACOB STRAIGHT-LINE METHOD

For sufficiently large values of t, or small values of r, Cooper and Jacob (1946) noticed that the infinite series for the well function (Equation 7.7) can be approximated by truncating the series after two terms. Thus, based on their approximation, the well function can be written as

$$W(u) \cong -0.5772 - \ln u = \ln\frac{0.561}{u} = \ln\left[2.25\frac{T}{S}\left(\frac{t}{r^2}\right)\right] \tag{7.10}$$

Using this approximation, the drawdown can be found from the following equation:

$$s(r,t) = \frac{Q}{4\pi T}W(u) = \frac{2.30Q}{4\pi T}\log\left[2.25\frac{T}{S}\left(\frac{t}{r^2}\right)\right] \tag{7.11a}$$

$$s(r,t) = \frac{2.30Q}{4\pi T}\log\left[2.25\frac{T}{S}\right] + \frac{2.30Q}{4\pi T}\log\left(\frac{t}{r^2}\right) \tag{7.11b}$$

The first term on the right-hand side of the previous equation is constant for an idealized Theis aquifer with a fully penetrating discharge well, operating at a constant Q. The variables, $s(r, t)$ and $\log(t/r^2)$, in Equation 7.11b, therefore, represent the equation of a straight line. Thus, Equation 7.11b should plot as a straight line on a semilogarithmic paper, with $s(r, t)$ plotted on the arithmetic scale and the variable t/r^2 plotted on the logarithmic scale (see Figure 7.6). The slope of the straight line in Figure 7.6, therefore, represents the quantity, $2.30\ Q/4\pi T$. The slope of the straight line on a semilogarithmic graph paper is simply equal to the drawdown change, Δs, per log cycle, as shown in Figure 7.6. It may be emphasized that the slope of the straight line in this figure is positive, because the increasing drawdown is plotted downward. Thus,

$$\Delta s = (s_2 - s_1) = \frac{2.30Q}{4\pi T} \tag{7.12a}$$

$$T = \frac{2.30Q}{4\pi(s_2 - s_1)} = \frac{2.30Q}{4\pi\Delta s} \tag{7.12b}$$

If the constant pumping rate, Q, and the slope of straight line, Δs, are known, the previous equation provides an estimate of transmissivity, T, of the idealized Theis aquifer.

To estimate the storativity, S, of the aquifer, we proceed as follows. Let us extrapolate the straight line in Figure 7.6, until it intersects with the

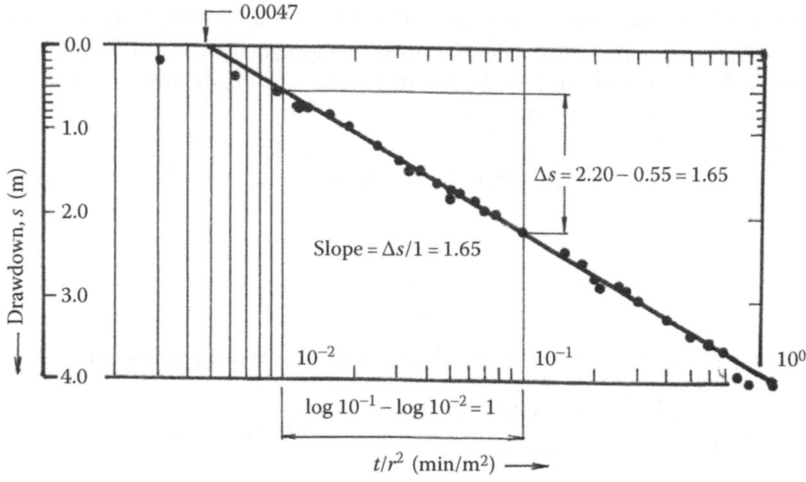

Figure 7.6 Semilogarithmic plot for the Jacob straight-line method.

zero-drawdown axis at some point with a positive value of abscissa t/r^2, as shown in the figure. Let this value be denoted by $(t/r^2)_0$. Since drawdown, $s(r, t)$, vanishes at $t/r^2 = (t/r^2)_0$, it follows that the well function, $W(u)$, should also become zero at $t/r^2 = (t/r^2)_0$. In other words, argument of the logarithmic function must become unity in Equation 7.10 at $t/r^2 = (t/r^2)_0$. Thus, setting the argument of the logarithmic function equal to unity provides the necessary equation for S as shown in the following:

$$S = 2.25T\left(\frac{t}{r^2}\right)_0$$

where the quantity $(t/r^2)_0$ is obtained graphically and T is obtained from Equation 7.12b. The details of the Jacob straight-line method are further elucidated, using *Illustrative Problem 7.4*.

7.4.1 Illustrative problem 7.4

For the field data shown in the table in Illustrative Problem 7.3, evaluate the aquifer constant using Jacob's straight-line method.

Solution: As before, we decide to use the minute as the unit for measuring time. The actual choice is arbitrary, but we must use a consistent set of units for measurements. Thus, the constant discharge

$$Q = 500 \text{ m}^3/\text{day} = 0.3472 \text{ m}^2/\text{min}.$$

To proceed further, we need to plot the data on a semilogarithmic graph paper. From this plot, we visually infer the best-fit straight line passing through the data. This step is shown in Figure 7.6. From this figure, we can obtain the following information:

Slope of best-fit straight line, $\Delta s = (s_2 - s_1) = 2.20 - 0.55 = 1.65$ m.

From Figure 7.6, the intercept of the line with the zero-drawdown axis is found as shown here:

$$\left(\frac{t}{r^2}\right)_0 = 0.0047 \text{ min/m}^2$$

The transmissivity can be obtained from the following expression:

$$T = \frac{2.30\,Q}{4\pi\Delta s} = \frac{2.30 \times 0.3472\,\text{m}^3/\text{min}}{4\pi \times 1.65\,\text{m}} = 0.0385\,\text{m}^2/\text{min}$$

The storativity can be obtained from the following expression (Equation 7.13):

$$S = 2.25T\left(\frac{t}{r^2}\right)_0$$

$$= 2.25 \times \left(0.0385\text{m}^2/\text{min}\right) \times \left(0.0047\text{min/m}^2\right) = 0.0004$$

Comparing the values of *T* and *S* with the previously obtained values using the Theis method, we observe slight differences, which can be attributed to the graphical nature of the solutions. Despite these comments, the two methods give reasonably close answer to transmissivity and storativity of the aquifer.

7.5 MODIFICATION OF THE JACOB METHOD: DISTANCE–DRAWDOWN METHOD

When simultaneous observations are made in three or more than three observation wells, a straightforward modification of the Jacob method is possible. In this case, the time of observation, $t = t_{obs}$, becomes a constant and Equation 7.11b can be modified to the following form:

$$s(r,t) = \frac{2.30\,Q}{4\pi T}\log\left[2.25\frac{Tt_{obs}}{S}\right] - \frac{2.30\,Q}{2\pi T}\log(r) \tag{7.13}$$

Since the first term on the right-hand side is constant, the aforementioned equation represents a straight line on the semilogarithmic graph paper, when *s*(*r*, *t*) is plotted on the arithmetic scale, and the radial distance *r*

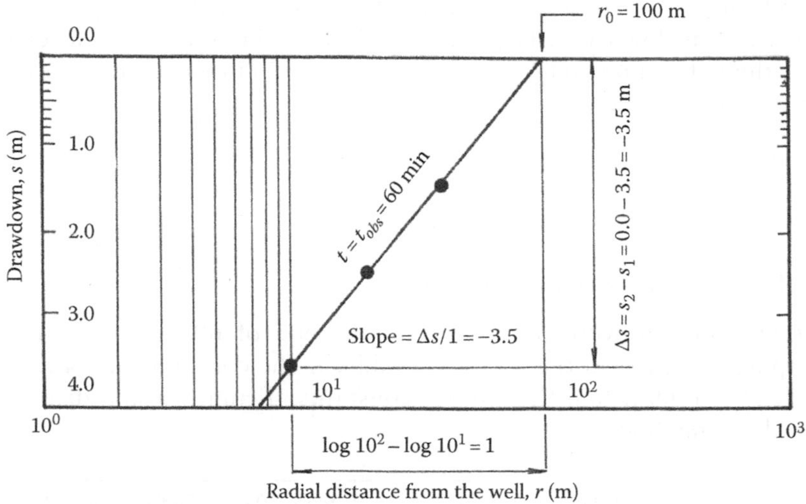

Figure 7.7 Modification of the Jacob method. Simultaneous drawdowns are plotted on the arithmetic scale against the radial distances plotted on the logarithmic scale.

on the logarithmic scale. The slope of this line represents the quantity, $-(2.30Q/2\pi T)$. On the semilogarithmic graph paper, the slope of the line is given by the following expression (Figure 7.7):

$$\text{Slope} = \frac{s_2 - s_1}{\log r_2 - \log r_1} = \frac{s_2 - s_1}{1} = \Delta s \tag{7.14}$$

where s_2 and s_1 are drawdowns, respectively, at radial distances r_2 and r_1. The second equality in the aforementioned expression is true if $r_2 > r_1$ and these radial distances are a log-cycle apart. Thus, equating the slope with the expression $-(2.30\,Q/2\pi T)$ yields the required equation for transmissivity as shown here:

$$T = -\frac{2.30\,Q}{2\pi(s_2 - s_1)} = -\frac{2.30\,Q}{2\pi\Delta s} \tag{7.15}$$

To compute storativity, we proceed similarly to the procedure followed in the case of the Jacob straight-line method. Using Cooper–Jacob's approximation, the drawdown at any time and at any radial distance can be obtained by the following equation:

$$s(r,t) = \frac{2.30\,Q}{4\pi T} \log\left[2.25\frac{T}{S}\left(\frac{t}{r^2}\right)\right] \tag{7.11a}$$

From Figure 7.7, we observe that, according to Cooper–Jacob's approximation, the drawdown vanishes at $r = r_0$. Thus, setting the argument of the logarithmic function equal to one in Equation 7.11a yields the necessary expression for S as shown in the following:

$$S = 2.25T\left(\frac{t_{obs}}{r_0^2}\right)$$

7.5.1 Illustrative problem 7.5

Instantaneous drawdown readings at the end of 60 min of constant pumping are given in the following table. If the constant pumping rate is $Q = 500\ \text{m}^3/\text{day}$, find the aquifer constants, T and S, using distance–drawdown method.

Time since pumping (min)	*Radial distance 10 m*	*Radial distance 20 m*	*Radial distance 40 m*
	Drawdown (m)	*Drawdown (m)*	*Drawdown (m)*
60	3.50	2.43	1.43

Solution: The distance–drawdown data are plotted on a semilogarithmic graph paper as shown in Figure 7.7. From this figure, the slope of the best-fit line is obtained as follows:

$$\text{Slope} = (s_2 - s_1)/1 = (0.0 - 3.5) = -3.5\ \text{m}.$$

The next step of the procedure is to find graphically, by extrapolating the straight line, the value of the radial distance r_0, where the drawdown is zero. From Figure 7.7, the value $r_0 = 100$ m is found. The transmissivity of the aquifer is found from Equation 7.15 as shown in the following:

$$T = \frac{-2.30Q}{2\pi\Delta s} = \frac{-2.3 \times 0.472\ \text{m}^3/\text{min}}{2\pi(-3.5\ \text{m})} = 0.036\ \text{m}^2/\text{min}$$

The storativity is found as

$$S = \frac{2.25Tt_{obs}}{r_0^2} = \left(2.25 \times 0.036\ \text{m}^2/\text{min}\right) \times \frac{60\ \text{min}}{10{,}000\ \text{m}^2} = 0.00049$$

7.6 REMARKS ON THE USE OF THE THIEM EQUATION IN THE CASE OF UNSTEADY FLOW CONDITION

The Thiem equation is based on the physics of steady-state (equilibrium) flow condition. As pointed out by Todd (1959, p. 82), *the equilibrium flow condition in an extensive aquifer is theoretically not possible.* From the transient analysis of Theis, it is also evident that the drawdown, $s(r, t)$, at a definite r keeps on increasing ad infinitum with time, t, and, more importantly, the drawdown curve becomes undefined as time approaches infinity. Thus, it is natural to suspect the estimate of transmissivity, T, based on the Thiem equation (7.1b). However, the empirical evidence is contrary to this suspicion. Values of transmissivity obtained by Equation 7.1b, even under nonequilibrium conditions, appear to be satisfactory. This fact is also recognized in the literature previously (Todd, 1959; Fetter, 2001).

The explanation of this paradox can be found in the works of Charbeneau (2000) and Hermance (1999). According to Charbeneau, the time-rate of change of drawdown after sufficiently long duration of pumping, or small values of r, is given by the following equation (Charbeneau, 2000, p. 108):

$$\frac{\partial s}{\partial t} \cong \frac{Q}{4\pi T}\left(\frac{1}{t}\right) \tag{7.16}$$

With respect to the aforementioned equation, there are two facts worth mentioning: (1) $\partial s/\partial t \to 0$, as $t \to \infty$ and (2) the time-rate of change of drawdown $\partial s/\partial t$ does not depend on the radial distance, r. In other words, the changes in drawdown, δs, is the same at all radial distances, during any increment of time δt. This observation is in concurrence with the findings of Hermance (see Figure 10.4, Hermance, 1999, p. 122). Both of these studies imply that, after an initial pumping, the *gradient* of phreatic surface stabilizes, despite of the fact that the drawdown itself docs not. In other words, after initial pumping, the difference in phreatic elevations $\left[\phi(r_2)-\phi(r_1)\right]$ at two arbitrary locations becomes invariant of time. Thus, in computing T from Equation 7.1b, it does not materially matter whether the difference $\left[\phi(r_2)-\phi(r_1)\right]$ is observed at a finite time or at a time which tends to infinity. The preceding argument is true provided the time of observation is sufficiently delayed since the beginning of pumping.

In this chapter, we have conveyed the basic idea behind the aquifer testing. The discussion has been directed toward students who are interested in finding the basic thought, without getting into cumbersome details of actual implementation of the method to field situations. For this purpose, the aquifer is simplified to represent the idealized Theis aquifer. We are all aware of the fact that the real situation in the field may be far from this idealization. The various treatments of the problem when the aquifer departs from this idealized situation have been given in the literature. The subject does not belong to an introductory textbook. The interested readers may, however, consult more comprehensive treatise on this subject. For instance, Batu (1998) treats the subject of aquifer hydraulics in greater details from the perspective of a practicing engineer.

7.7 EXERCISES

7.1 A fully penetrating well in a confined aquifer is pumped at a constant rate, 4000 m^3/day. Assuming a homogeneous, isotropic, horizontal confined aquifer with $T=400$ m^2/day and $S=0.0005$, find drawdowns at distances 10 and 15 m from the well, after 2 days of pumping. For the solution, use Jacob's approximation for the well function.

7.2 Redo Exercise 7.1 using $T=200$ m^2/day. All other data remain the same as in the enunciation of Exercise 7.1.

7.3 By comparing the results obtained in Exercises 7.1 and 7.2, discuss qualitatively the influence of transmissivity on the potentiometric (piezometric) surface.

7.4 During an aquifer testing, the time–drawdown data were obtained as shown in the following table. Using the Theis method, determine the aquifer constants, T and S, if the pumping rate, $Q=1000$ m^3/day, were kept constant during the aquifer testing.

Time since pumping (min)	*Radial distance 10 m*	*Radial distance 20 m*	*Radial distance 40 m*
	Drawdown (m)	*Drawdown (m)*	*Drawdown (m)*
5	2.07	1.27	0.32
10	2.87	1.59	0.79
20	3.18	2.23	1.51
30	3.62	2.47	1.59
60	4.30	3.18	2.07
90	4.70	3.18	2.39
120	4.93	3.58	2.47

7.5 For the data given in Exercise 7.4, evaluate the aquifer constant using Jacob's straight-line method.

7.6 Using distance–drawdown method, find the aquifer constants (T and S) utilizing the simultaneous drawdown data at 60, 90, and 120 min after pumping of well at a constant rate, $Q=1000$ m^3/day, as given in Exercise 7.4. Should the answer, theoretically speaking, be the same in three cases? If they are same, find graphically, as well as analytically, the relationship between radius of influence, r_0, and time, t. If they are not same, find the average values of T and S.

7.7 Aquifer testing based on the Theis method was performed in a confined aquifer of depth 20 m with a fully penetrating well discharging at a constant rate $Q=1200$ m^3/day. After finding the best fit between the data plot and the reverse type curve, the following coordinates of the match point were found: $W_p=1$, $(1/u)_p=1$ on the type curve, and $s_p=0.8$ m, $(t/r^2)_p=0.003$ min/m^2 on the data plot. Find the aquifer constants T and S.

SUGGESTED READINGS

Batu, V. 1998. *Aquifer Hydraulics: A Comprehensive Guide to Hydrogeologic Data Analysis*, John Wiley & Sons, Inc., New York.

Bear, J. 1988. *Dynamics of Fluids in Porous Media*, Dover Publications, Inc., New York.

Box, M. J. April 1965. A new method of constrained optimization and comparison with other methods, *Computer Journal*, 8, 42–52.

Charbeneau, R. J. 2000. *Groundwater Hydraulics and Pollutant Transport*, Prentice Hall, Inc., Upper Saddle River, NJ.

Cooper, H. H. Jr. and C. E. Jacob. 1946. A generalized graphical method for evaluating formation constants and summarizing well-field history, *American Geophysical Union Transactions*, 27, 526–534.

Fetter, C. W. 2001. *Applied Hydrogeology*, 4th edn., Prentice Hall, Inc., Upper Saddle River, NJ.

Freeze, R. A. and J. A. Cherry. 1979. *Groundwater*, Prentice Hall, Englewood Cliffs, NJ.

Haque, M. I. 1985. Optimal design of plane frames by the complex method, in *Advances and Trends in Structures and Dynamics*, eds. A. K. Noor and R. J. Hayduk, Pergamon Press, New York, pp. 451–456.

Hermance, J. F. 1999. *A Mathematical Primer on Groundwater Flow*, Prentice Hall, Upper Saddle River, NJ.

Jacob, C. E. 1950. Flow of ground-water, in *Engineering Hydraulics*, ed. H. Rouse, John Wiley & Sons, New York, pp. 321–386.

Strack, O. D. L. 1989. *Groundwater Mechanics*, Prentice Hall, Inc., Englewood Cliffs, NJ.

Theis, C. V. 1935. The relationship between the lowering of the piezometric surface and the rate and discharge of a well using ground-water storage, *American Geophysical Union Transactions*, 2, 519–524.

Todd, D. K. 1959. *Ground Water Hydrology*, John Wiley & Sons, Inc., New York.

Wenzel, L. K. 1942. *Method of Determining Permeability of Water-Bearing Materials with Special Reference to Discharging Well Methods*, U.S. Geological Survey Water Supply Paper 887.

Chapter 8

Coastal aquifers

In this chapter, we shall discuss the flow of groundwater in the vicinity of coastlines. For this purpose, we shall assume that the interface between the saltwater and the freshwater is sharp without molecular diffusion or mechanical dispersion. Figure 8.1 illustrates schematically the various possibilities of the groundwater flow toward the actual coastline with and without a sharp interface. To simplify the analysis, we shall make throughout this chapter the following assumptions: (1) the interface between the freshwater and saltwater is sharp; (2) the saltwater is in a stagnant state; and (3) the freshwater migrates toward the coastline, and this migration may be analyzed by using the Dupuit–Forchheimer assumptions.

8.1 GHYBEN–HERZBERG PRINCIPLE

Studies in the late nineteenth century by W. Baydon-Ghyben and A. Herzberg independently lead to the following equation, describing the interface between the freshwater and saltwater of the seas (Figure 8.2):

$$h_s(x,y) = \frac{\rho_f}{\rho_s - \rho_f} h_f(x,y) \qquad (8.1)$$

where

ρ denotes the density of the fluid
h denotes the height as shown in Figure 8.2
Subscripts f and s refer to freshwater and saltwater, respectively

Equation 8.1 is strictly valid if the interface between the freshwater and saltwater is *sharp* and the fluids are in hydrostatic equilibrium.

It is apparent from Figure 8.2 that, in the case of Ghyben–Herzberg analysis, the discharge of freshwater into the sea takes place along a line with zero discharge area. As pointed out by Hubbert, the Ghyben–Herzberg analysis must break down near the shoreline in order to provide a seepage

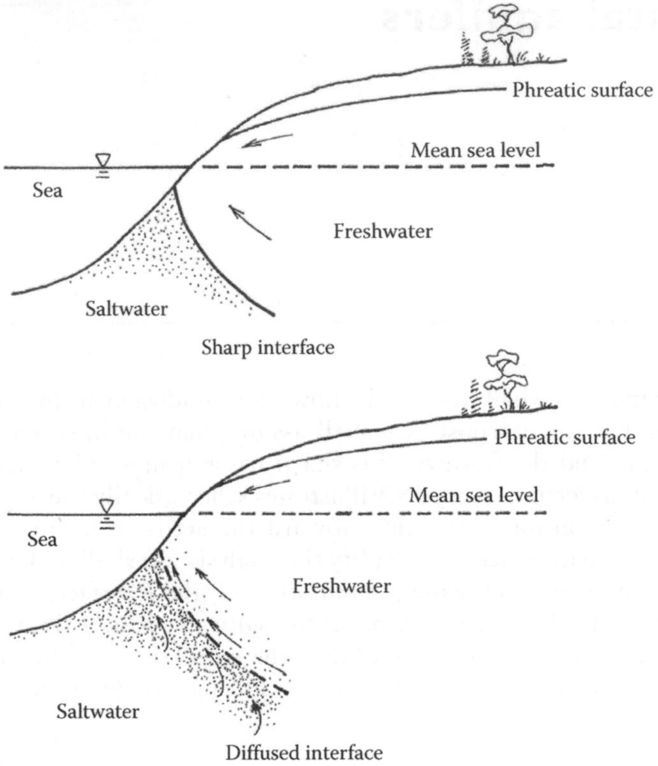

Figure 8.1 Schematic cross sections of coastal aquifers.

face for the outflow of freshwater. A more realistic picture of groundwater flow near the shore is shown in Figure 8.3, which has been adapted after Hubbert's work (1940).

8.2 STRACK'S ANALYSIS: INSTABILITY CAUSED BY A FULLY PENETRATING WELL IN A SHALLOW COASTAL AQUIFER

Following Strack, we assume a priori the following general linear relationship between h and ϕ:

$$h = \alpha\phi + \beta \tag{8.2}$$

where

α and β are arbitrary constants

The height h is defined in Figure 8.4, along with the other pertinent variables

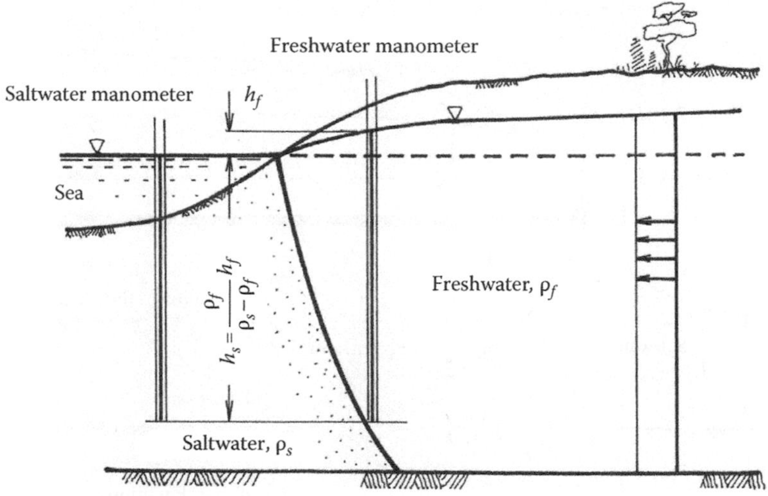

Figure 8.2 Definition sketch for Ghyben–Herzberg equation.

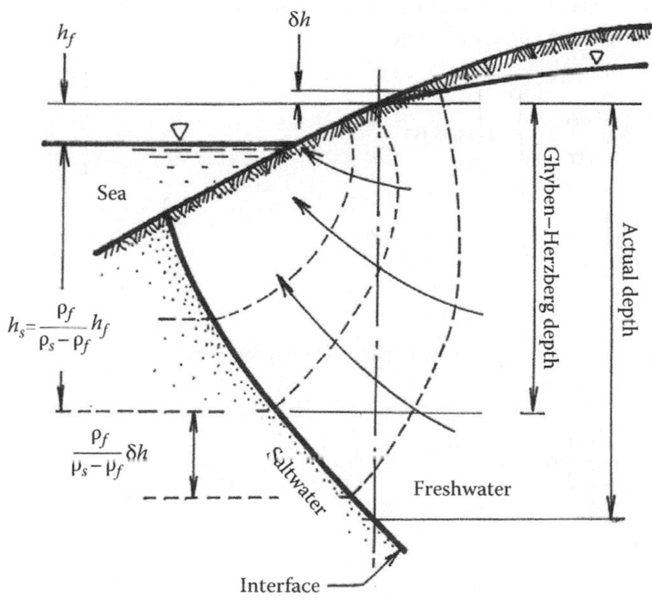

Figure 8.3 Equipotentials and flow lines near the shore. (After Hubbert, M.K., *J. Geol.*, 48, 785, 1940.)

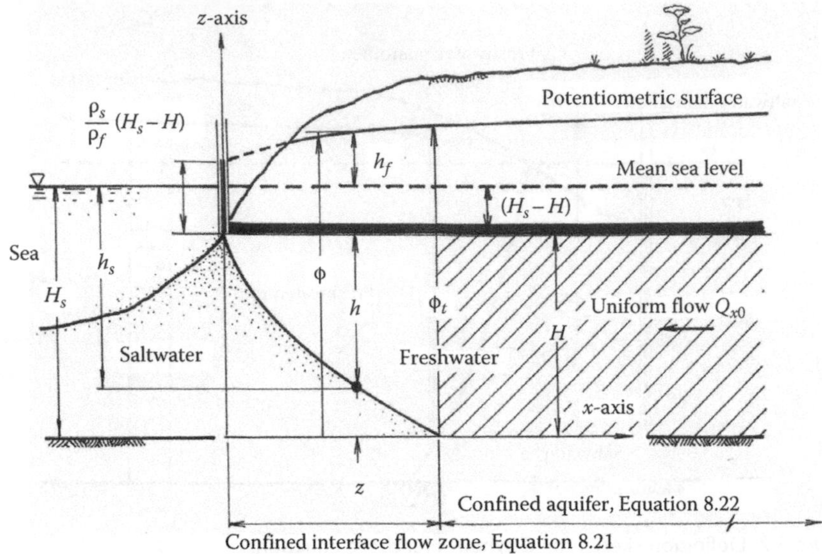

Figure 8.4 Nomenclature for shallow confined interface flow region and shallow confined aquifer.

Using Darcy's law for homogeneous isotropic aquifer, we obtain the following expressions for the discharge:

$$Q_x = -Kh\frac{\partial \phi}{\partial x} = -K(\alpha\phi + \beta)\frac{\partial \phi}{\partial x} \tag{8.3a}$$

$$Q_y = -Kh\frac{\partial \phi}{\partial y} = -K(\alpha\phi + \beta)\frac{\partial \phi}{\partial y} \tag{8.3b}$$

The aforementioned equation for $\alpha \neq 0$ can be written as

$$Q_x = -\frac{\partial}{\partial x}\left[\frac{1}{2}\alpha K\left(\phi + \frac{\beta}{\alpha}\right)^2 + C\right] \tag{8.4a}$$

$$Q_y = -\frac{\partial}{\partial y}\left[\frac{1}{2}\alpha K\left(\phi + \frac{\beta}{\alpha}\right)^2 + C\right] \tag{8.4b}$$

Thus, for $\alpha \neq 0$, the discharge potential becomes

$$\Phi = \left[\frac{1}{2}\alpha K\left(\phi + \frac{\beta}{\alpha}\right)^2 + C\right] \tag{8.5}$$

If $\alpha = 0$, it follows from Equation 8.2 and Darcy's law that the following relationships are true:

$$Q_x = -Kh\frac{\partial \phi}{\partial x} = -K\beta\frac{\partial \phi}{\partial x} = -\frac{\partial(K\beta\phi + C)}{\partial x} \tag{8.6a}$$

$$Q_y = -Kh\frac{\partial \phi}{\partial y} = -K\beta\frac{\partial \phi}{\partial y} = -\frac{\partial(K\beta\phi + C)}{\partial y} \tag{8.6b}$$

which in turn implies that in the case of $\alpha = 0$ the discharge potential is given by the following:

$$\Phi = K\beta\phi + C \tag{8.7}$$

In Equations 8.2 through 8.7, ϕ denotes the piezometric head with respect to a horizontal impervious bedrock and C, an arbitrary constant. However, in all cases—whether alpha (α) is equal to zero or not—the discharge potential Φ is indeed a potential function, and it satisfies the Laplace equation (see Sections 4.1 and 4.2).

8.2.1 Shallow confined interface flow region

The geometry of this region is defined in Figure 8.4. Our objective in this section is to establish the relationship between the flow depth, h, and the piezometric head, ϕ, in a shallow confined interface flow region. For this purpose, we shall assume that the sea level H_s, the depth of confined aquifer H, and the densities of freshwater and saltwater, ρ_f and ρ_s, are known data. The other variables such as h, h_f, and ϕ are considered as the unknown. From Figure 8.4, the following two geometrical relationships can be easily established:

$$h = h_s - (H_s - H) \tag{8.8}$$

$$\phi = H_s + h_f \tag{8.9}$$

From Ghyben–Herzberg equation, the following relationship between the unknowns can also be determined:

$$h_f = \frac{\rho_s - \rho_f}{\rho_f} h_s \tag{8.10}$$

By substituting the value of h_s from Equation 8.10 into Equation 8.8, we obtain the following:

$$h = \frac{\rho_f}{\rho_s - \rho_f} h_f - (H_s - H) = \frac{\rho_f}{\rho_s - \rho_f}(\phi - H_s) - (H_s - H) \tag{8.11}$$

The second equality in the aforementioned equation follows from the geometric relationship $h_f = \phi - H_s$, given in Equation 8.9. Equation 8.11 can be expressed as

$$h = \left[\frac{\rho_f}{\rho_s - \rho_f}\right]\phi - \left[\frac{\rho_s}{\rho_s - \rho_f}\right]H_s + H \tag{8.12}$$

It is now clear that the relationship between h and ϕ is linear in the case of the shallow confined interface flow region. Comparing the aforementioned equation with Equation 8.2, we obtain

$$\alpha = \frac{\rho_f}{\rho_s - \rho_f} \tag{8.13a}$$

$$\beta = -\left[\frac{\rho_s}{\rho_s - \rho_f}\right]H_s + H \tag{8.13b}$$

$$\frac{\beta}{\alpha} = \frac{-\rho_s H_s + (\rho_s - \rho_f)H}{\rho_f} \tag{8.13c}$$

Thus, flows through the shallow confined interface region can be described by the discharge potential:

$$\Phi = \frac{K}{2}\left(\frac{\rho_f}{\rho_s - \rho_f}\right)\left[\phi - \frac{\rho_s H_s - (\rho_s - \rho_f)H}{\rho_f}\right]^2 + C_{ci} \tag{8.14}$$

where

C_{ci} represents an arbitrary constant

The subscript *ci* refers to the confined interface flow region

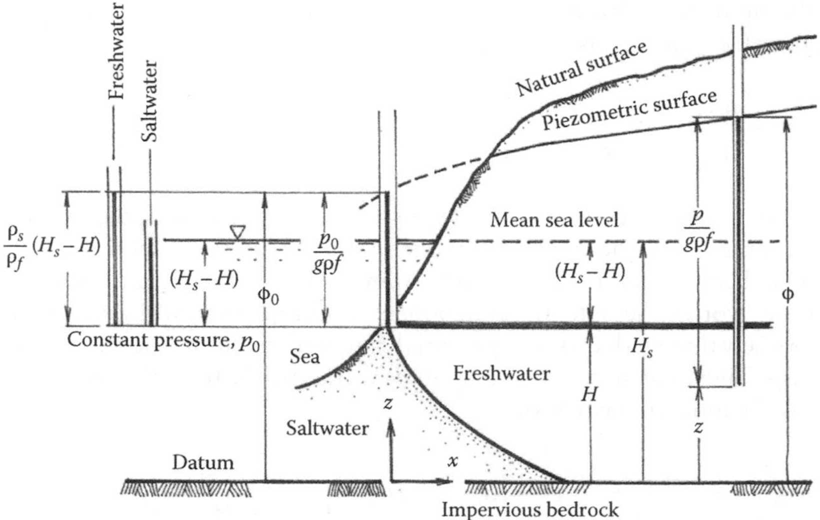

Figure 8.5 Piezometric head at the coastline ($x = 0$).

The second term inside the square brackets represents the piezometric head, $\phi = \phi_0$, at $x = 0$, as shown in Figure 8.5. From this figure, it can be readily seen that

$$\phi_0 = H + \frac{\rho_s}{\rho_f}(H_s - H) = \frac{\rho_s H_s - (\rho_s - \rho_f) H}{\rho_f} \tag{8.15}$$

Thus, Equation 8.14 can be written as

$$\Phi = \frac{K}{2}\left(\frac{\rho_f}{\rho_s - \rho_f}\right)[\phi - \phi_0]^2 + C_{ci} \tag{8.16}$$

8.2.2 Shallow confined flow region and continuity of discharge potential

The discharge potential for the shallow confined flow region is known (Chapter 4):

$$\Phi = KH\phi - \frac{KH^2}{2} \tag{8.17}$$

At the interregional boundary, the discharge potentials given by Equations 8.17 and 8.14 should be continuous. This interregional boundary is located at the tip of the saltwater tongue, as shown in Figure 8.4. At the tip, the piezometric head, ϕ_t, is given by the following expression:

$$\phi_t = \frac{\rho_s}{\rho_f} H_s \tag{8.18}$$

In order to ensure continuity of the discharge potential Φ at the tip of the saltwater tongue, the discharge potential must have the same value at the tip whether this point is approached from the left or the right in the figure. In other words, the potential function given in Equation 8.14 must acquire the same value as that of the potential function given in Equation 8.17 at the tip. Thus, equating these potentials at the tip of the saltwater tongue yields the following equation:

$$\frac{K}{2}\left(\frac{\rho_f}{\rho_s - \rho_f}\right)\left[\phi_t - \frac{\rho_s H_s - (\rho_s - \rho_f) H}{\rho_f}\right]^2 + C_{ci} = KH\phi_t - \frac{KH^2}{2} \tag{8.19}$$

where the piezometric head at the tip ϕ_t is given by Equation 8.18. By substituting the value of ϕ_t from Equation 8.18 into Equation 8.19 and solving for the arbitrary constant C_{ci}, one obtains the following result:

$$C_{ci} = \frac{K}{2}\frac{\rho_s}{\rho_f}\left[2HH_s - H^2\right] \tag{8.20}$$

The constant C_{ci} is no more arbitrary, it must maintain the aforementioned value in order for the discharge potential to be continuous across the tip of the saltwater tongue.

8.2.3 Flow prior to pumping of well

We assume there is a unidirectional flow in the negative x-axis through the combined aquifer region, comprising of a *shallow confined interface flow region* and a *shallow confined aquifer* as shown in Figure 8.4. This flow is described by a continuous discharge potential Φ, which satisfies the Laplace equation in the entire flow region. In the *shallow confined interface flow region*, this potential function is described by the following mathematical expression:

$$\Phi = \frac{K}{2}\left(\frac{\rho_f}{\rho_s - \rho_f}\right)\left[\phi - \phi_0\right]^2 + \frac{K}{2}\frac{\rho_s}{\rho_f}\left[2HH_s - H^2\right] \tag{8.21}$$

and in the *shallow confined flow region*, by the following mathematical expression:

$$\Phi = KH\phi - \frac{KH^2}{2} \tag{8.22}$$

Despite two different mathematical expressions for its description, Φ remains a continuous harmonic function that satisfies the Laplace equation $\nabla^2\Phi = 0$ in the entire flow region. Since the flow problem is unidirectional along the negative x-axis, the general solution of the Laplace equation is given by the following:

$$\Phi = Ax + B \tag{8.23}$$

The constants A and B can be found from the following boundary conditions:

$$\Phi = \Phi_0 \quad \text{at } x = 0 \tag{8.24a}$$

$$\Phi = \Phi_1 \quad \text{at } x = L \tag{8.24b}$$

Substituting the values of A and B, obtained by boundary conditions (8.24a) and (8.24b), Equation 8.23 can be written as

$$\Phi = \frac{\Phi_1 - \Phi_0}{L}x + \Phi_0 = -Q_{x0}x + \Phi_0 \tag{8.25}$$

where Q_{x0} denotes the discharge per width through the entire depth of the confined aquifer. The negative sign in the aforementioned equation implies that the flow through the aquifer is in the negative x-direction (Figure 8.4).

8.2.4 Discharge potential for combined flow

In the absence of the well, the discharge potential for unidirectional flow toward the coast is given by Equation 8.25. The discharge potential for flow toward a well, located at a distance d from the coastline, can be obtained by the method of images. This discharge potential due to a well in the vicinity of the coastline is (see Chapter 4)

$$\Phi = \frac{Q}{2\pi}\ln\frac{r_1}{r_2} = \frac{Q}{4\pi}\ln\frac{(x-d)^2 + y^2}{(x+d)^2 + y^2} \tag{8.26}$$

The aforementioned discharge potential is based on the choice that $\Phi = 0$ at the coast. The combined flow field is described by the combination of the two potentials given in Equations 8.25 and 8.26. Thus, the discharge potential for the combined flow field becomes

$$\Phi = -Q_{x0}x + \frac{Q}{4\pi}\ln\frac{(x-d)^2+y^2}{(x+d)^2+y^2} + \Phi_0 \tag{8.27a}$$

$$\Phi(x,y) = -Q_{x0}x + \frac{Q}{4\pi}\ln\frac{(x-d)^2+y^2}{(x+d)^2+y^2} + \frac{K}{2}\frac{\rho_s}{\rho_f}\left[2HH_s - H^2\right] \tag{8.27b}$$

The last equation describes the discharge potential at a generic point (x, y) in the combined flow field, due to a well pumping in the vicinity of the coastline in the presence of a unidirectional flow toward the coastline.

8.2.5 Shape of the tip of the saltwater tongue

The value of piezometric head at the tip has been given previously as (Figure 8.4)

$$\phi_t = \frac{\rho_s}{\rho_f}H_s \tag{8.18}$$

Substitution of this value into Equation 8.22 yields the following expression for the discharge potential at the tip:

$$\Phi_t = KH\phi_t - \frac{KH^2}{2} = K\frac{\rho_s}{\rho_f}HH_s - \frac{KH^2}{2} \tag{8.28}$$

It is evident from this equation that for a given problem the discharge potential remains constant along the tip of the saltwater tongue. Thus, the shape of the tip in x, y-plane can be ascertained by the substitution of this constant into Equation 8.27b, as shown in the following:

$$K\frac{\rho_s}{\rho_f}HH_s - \frac{KH^2}{2} = -Q_{x0}x + \frac{Q}{4\pi}\ln\frac{(x-d)^2+y^2}{(x+d)^2+y^2} + \frac{K}{2}\frac{\rho_s}{\rho_f}\left[2HH_s - H^2\right] \tag{8.29}$$

which can be written in the following dimensionless form:

$$\frac{-KH^2}{Q_{x0}d} \times \frac{\rho_s - \rho_f}{\rho_f} = \frac{2x}{d} - \frac{Q}{Q_{x0}d} \times \frac{1}{2\pi}\ln\frac{\left(\frac{x}{d}-1\right)^2 + \left(\frac{y}{d}\right)^2}{\left(\frac{x}{d}+1\right)^2 + \left(\frac{y}{d}\right)^2} \tag{8.30}$$

Using the following notations for the dimensionless quantities

$$\lambda \equiv \frac{-KH^2}{Q_{x0}d} \times \frac{\rho_s - \rho_f}{\rho_f} \tag{8.31a}$$

$$\mu \equiv -\frac{Q}{Q_{x0}d} \tag{8.31b}$$

$$x' \equiv \frac{x}{d} \tag{8.31c}$$

$$y' \equiv \frac{y}{d} \tag{8.31d}$$

Equation 8.30 can be written as

$$\lambda = 2x' + \frac{\mu}{2\pi} \ln \frac{(x'-1)^2 + (y')^2}{(x'+1)^2 + (y')^2} \tag{8.32}$$

It may be emphasized that both λ and μ are positive quantities because Q_{x0} is negative.

8.2.6 Relationship between the location of the tip on the x-axis and the well discharge

The aforementioned equation describes the geometry of the tip of the saltwater tongue in terms of the dimensionless coordinates x' and y'. If we denote the location of the tip on the x-axis by $x_t'(=x_t/d)$, we can obtain an implicit expression for x_t' by substituting x_t' for x' and setting $y'=0$ in Equation 8.32. Thus, the following implicit expression describes the tip location on the x-axis:

$$\lambda = 2x_t' + \frac{\mu}{2\pi} \ln \frac{(x_t'-1)^2}{(x_t'+1)^2} \tag{8.33}$$

For a given problem, $\lambda \equiv \frac{-KH^2}{Q_{x0}d} \times \frac{\rho_s - \rho_f}{\rho_f}$ is a constant and $\mu \equiv -\frac{Q}{Q_{x0}d}$ represents the normalized well discharge. It is, therefore, evident that for a given problem (λ=*constant*), Equation 8.33 describes the location of the stable saltwater tongue tip (x_t') as a function of the normalized well discharge (μ). This functional relationship is graphically illustrated in Figure 8.6 for λ=0.5.

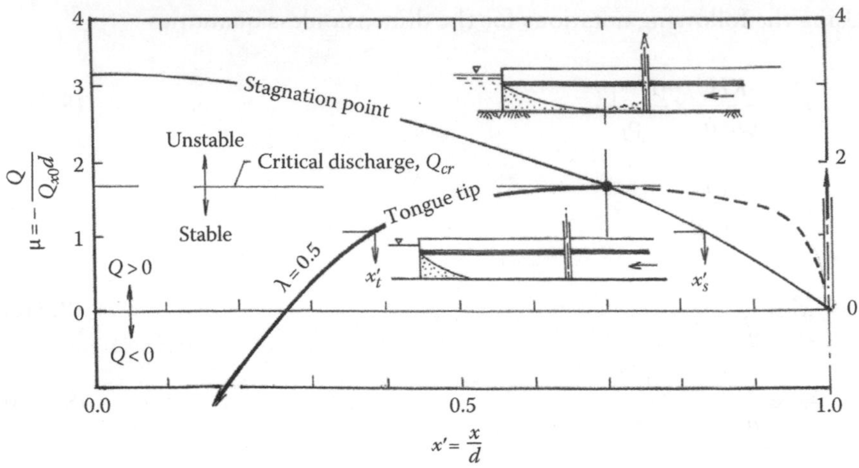

Figure 8.6 Relationship between tip location and well discharge for $\lambda=0.5$.

From this figure, it is clear that the relationship between x'_t and μ has a maxima. Thus, there is no possible *stable tip location* for well discharges above a certain maximum value. In other words, the maximum well discharge corresponding to the *maxima point* represents the *threshold* of instability. This condition is labeled in Figure 8.6 as the *critical well discharge*.

To further explore the behavior of the maxima point, Figure 8.7 has been prepared for two different values of $\lambda=0.3$ *and* $\lambda=0.5$. There are a number of points about this figure that deserve special attention. For instance, points identified by A and B represent two different given problems. In both cases, the *well discharge is zero* and the saltwater tongues are in *stable equilibrium* with the unidirectional flow Q_{x0} in the negative x-axis. The case A pertains to relatively lesser transmissivity value than that of case B. However, in both cases as the well discharge increases up to a maximum value, the tip of the saltwater tongue moves toward the well. Also, in both cases, the functional relationship is characterized by the presence of the maxima point.

Since the maxima point represents the critical point, it is natural to investigate the locus of this point. From a mathematical point of view, it is also possible to determine the locus of the maxima point by differentiating Equation 8.33 with respect to x'_t and setting the derivative of the normalized well discharge equal to zero. Such a differentiation yields the following results:

$$0 = 2 + \frac{1}{\pi}\ln\frac{(x'_t-1)}{(x'_t+1)}\frac{d\mu}{dx'_t} + \frac{\mu}{\pi}\left[\frac{1}{x'_t-1} - \frac{1}{x'_t+1}\right] \tag{8.34a}$$

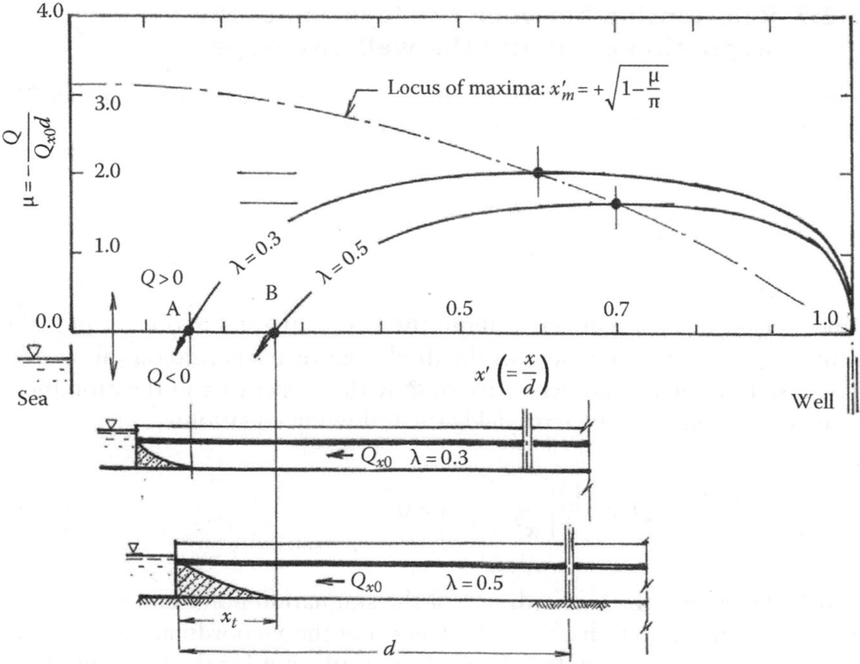

Figure 8.7 Locus of maxima point.

Let x'_m denote the abscissa of the maxima point. Then substitution of x'_m for x'_t in Equation 8.34a and remembering that $d\mu/dx'_t = 0$ at $x'_t = x'_m$ yields the following relationship for the location of the maxima point:

$$0 = 2\left[1 + \frac{\mu/\pi}{x'^2_m - 1}\right] \tag{8.34b}$$

or

$$x'_m = +\sqrt{1 - \frac{\mu}{\pi}} \tag{8.34c}$$

which shows that the maxima point is a function of dimensionless well discharge, μ, only. The aforementioned equation describes the locus of the maxima point in x', μ-plane. As we shall see in the subsequent section, the locus of the maxima point is identical to the locus of the stagnation point in x', μ-plane.

8.2.7 Relationship between the location of the stagnation point and the well discharge

The combined discharge potential along the x-axis is obtained by setting $y=0$ in Equation 8.27a as shown in the following:

$$\Phi(x,0) = -Q_{x0}x + \frac{Q}{4\pi}\ln\frac{(x-d)^2}{(x+d)^2} + \Phi_0 \tag{8.35}$$

Since the problem is symmetric about the x-axis, the stagnation point is the same where the x-component of the discharge vector is zero on the x-axis. Thus, setting the partial derivative of Φ with respect to x in the aforementioned equation equal to zero yields the following equation:

$$\frac{\partial\Phi}{\partial x} = \frac{d\Phi}{dx} = -Q_{x0} + \frac{Q}{\pi}\left[\frac{d}{x^2 - d^2}\right] = 0 \tag{8.36}$$

which shows that the x-coordinate of the stagnation point on the x-axis is a function of the well discharge Q. Denoting the x-coordinate of the stagnation point by x_s, the following explicit expression for the location of the stagnation point can be found:

$$\frac{x_s}{d} = +\sqrt{1 - \frac{\mu}{\pi}} \tag{8.37a}$$

which, in terms of the normalized x-coordinate, becomes

$$x_s' = +\sqrt{1 - \frac{\mu}{\pi}} \tag{8.37b}$$

where $\mu = -Q/Q_{x0}d$ is the normalized well discharge. Figure 8.8 has been prepared to illustrate the aforementioned relationship between x_s' and μ. From this figure, it is obvious that the stagnation point moves from the well toward the coastline as the well discharge increases. For $\mu=0$, the stagnation point is at the well and for $\mu=\pi$, it is at the coastline.

8.2.8 Mechanics of instability

Based on the instability theory presented by Strack, the saltwater wedge becomes unstable (or reaches the threshold of instability) when the location

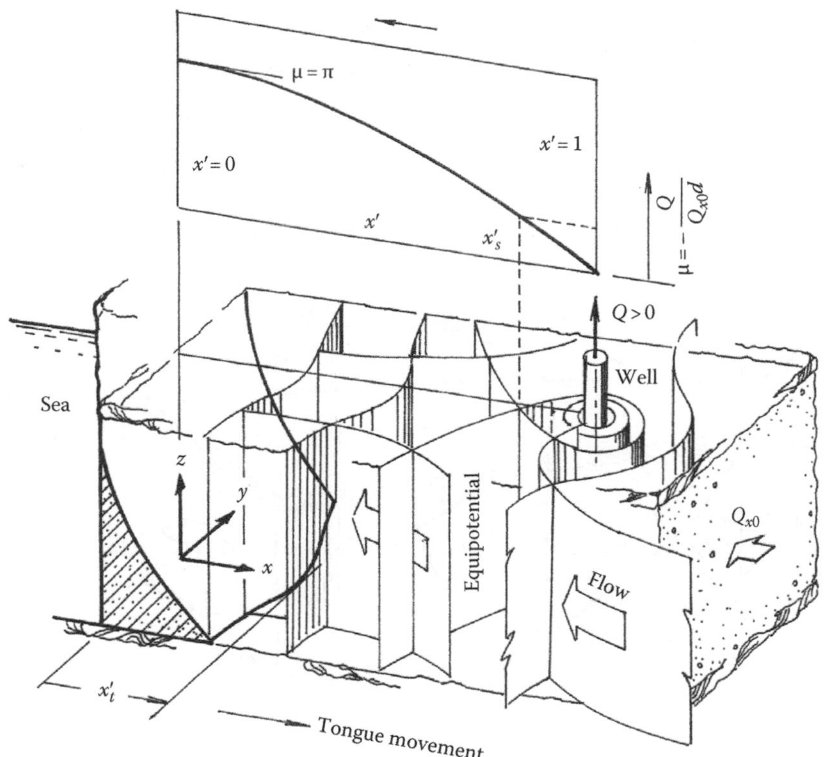

Figure 8.8 Location of stagnation point.

of the stagnation point coincides with the location of the tip of the saltwater tongue on the x-axis. This situation is schematically illustrated in Figure 8.9. Once the tip of the saltwater tongue moves toward the well from its location at the stagnation point, there is no stable equilibrium possible and the well discharges the mixed (saltwater and freshwater) flow. We can, therefore, express the *threshold condition* by requiring that $x_t' = x_s'$. Thus, substituting the value of x_s' from Equation 8.37b for x_t' in Equation 8.33 yields the following *critical condition*:

$$\lambda = 2\sqrt{1-\frac{\mu}{\pi}} + \frac{\mu}{\pi}\ln\frac{\left(1-\sqrt{1-\frac{\mu}{\pi}}\right)}{\left(1+\sqrt{1-\frac{\mu}{\pi}}\right)} \tag{8.38}$$

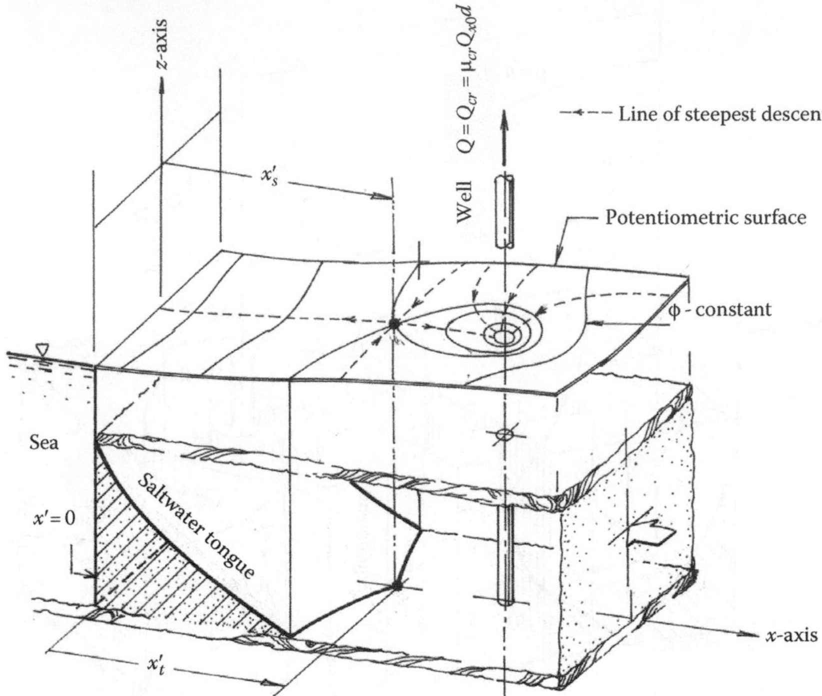

Figure 8.9 Illustration explaining the physics of incipient instability.

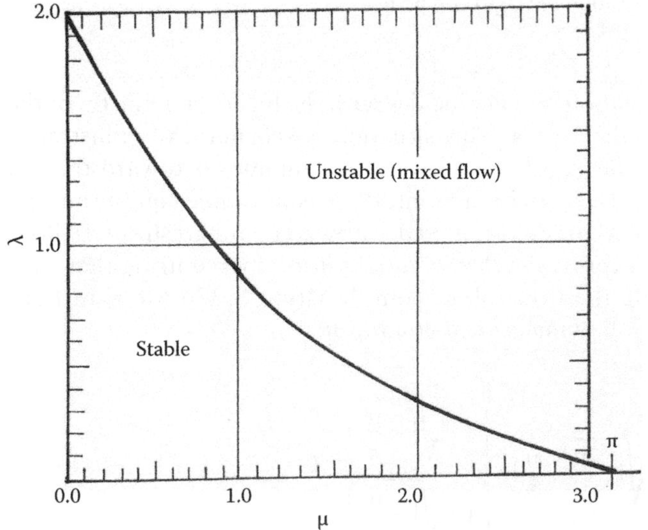

Figure 8.10 Stability chart.

Or the following inequality describes the *unstable* condition:

$$\lambda < 2\sqrt{1-\frac{\mu}{\pi}}+\frac{\mu}{\pi}\ln\frac{\left(1-\sqrt{1-\frac{\mu}{\pi}}\right)}{\left(1+\sqrt{1-\frac{\mu}{\pi}}\right)} \tag{8.39}$$

Equation 8.38 represents a curve in λ, μ-plane as shown in Figure 8.10. The area toward the right of the curve represents the unstable area.

SUGGESTED READINGS

Hubbert, M. K. 1940. The theory of ground-water motion, *Journal of Geology*, 48, 785–944.

Hubbert, M. K. 1957. *Darcy's Law and the Field Equations of the Flow of Underground Fluids*, Bulletin de l'Association Internationale d' Hydrologie Scientifique, n° 5, 1957. Also, Publication No. 104, Shell Development Company, Exploration and Production Research Division, Houston, TX.

Strack, O. D. L. 1989. *Groundwater Mechanics*, Prentice Hall, Inc., Englewood Cliffs, NJ.

Todd, D. K. 1959. *Ground Water Hydrology*, John Wiley & Sons, Inc., New York.

Chapter 9

Finite element method

In this chapter, we introduce the reader to the finite element method—one of the most flexible and powerful methods for numerical computations of engineering problems in general and of groundwater flows in particular. Our purpose here is twofold: one, to introduce the theory behind the numerical method quickly; and two, to prepare the reader for the implementation of the method using computers. For his purpose, a deliberate attempt is made to present the method in its simplest form. Although the method is quite general, we shall restrict it to the solution of the Laplace equation in the context of steady-state groundwater flows in a known region.

9.1 STEADY-STATE GROUNDWATER FLOW IN A KNOWN TWO-DIMENSIONAL REGION

9.1.1 Formulation of boundary-value problem

The steady flow of groundwater in a known two-dimensional flow region $\mathcal{R}$ can be formulated as the following boundary-value problem:

Find a function $\phi(x, y)$ such that

$$\frac{\partial^2\phi}{\partial x^2}+\frac{\partial^2\phi}{\partial y^2}=0 \quad \text{in } \mathcal{R} \tag{9.1a}$$

subject to the boundary conditions:

$$\phi=\overline{\phi} \quad \text{on } S_\phi \tag{9.1b}$$

$$\frac{\partial\phi}{\partial n}=\overline{q_n} \quad \text{on } S_q \tag{9.1c}$$

where ϕ and $\overline{q_n}$ denote the piezometric head and a specified function on S_q, respectively. The partial derivative, $\partial\phi/\partial n$, denotes the derivative along

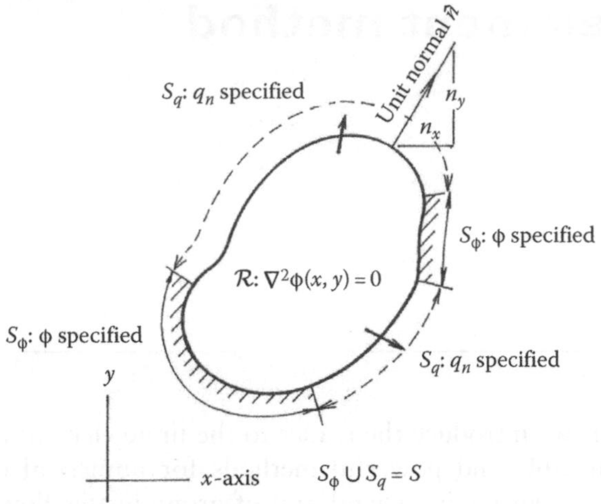

Figure 9.1 Definition sketch for boundary-value problem.

the normal to the surface S_q. The overscored quantities indicate the known functions. The boundaries S_ϕ and S_q represent the part of the boundary where ϕ and q_n are specified, respectively. The union of S_ϕ and S_q constitutes the entire boundary S (see Figure 9.1).

In finite element formulation, we do not directly work with the aforementioned boundary-value problem. Instead, we work with the corresponding *calculus of variations* formulation of the boundary-value problem. The discussion on what represents the corresponding calculus of variations formulation constitutes the subject matter of the following section.

9.1.2 Corresponding calculus of variations problem

We assume a priori the following functional defined for functions $\phi(x, y)$ meeting the requirement $\phi = \bar{\phi}$ on S_ϕ:

$$J(\phi) = \frac{1}{2} K \int_{\mathcal{R}} \left[\left(\frac{\partial \phi}{\partial x} \right)^2 + \left(\frac{\partial \phi}{\partial y} \right)^2 \right] d\mathcal{R} + \int_{S_q} \overline{q_n} \phi \, dS \tag{9.2}$$

With regard to the aforementioned equation, J is called the *functional* in the vernacular of calculus of variations. A functional is simply a function whose arguments are themselves functions defined over some region $\mathcal{R}$. The important fact from our perspective, however, is that the functional J attains a real value for each (given) function $\phi(x, y)$. On the very outset, we shall like to assert that the functional attains a stationary value (in fact, a minimum value) when $\phi(x, y)$ satisfies the aforementioned boundary-value problem—that is, $\phi(x, y)$ is a solution to the boundary-value problem. The proof of this assertion follows.

Let $\delta\phi$ denote an arbitrary change in function $\phi(x, y)$. As we shall see later, $\delta\phi$ is arbitrary but satisfies some restrictions imposed by mathematics. Using the jargons of calculus of variations, such an arbitrary change in the independent argument of J is called an *admissible variation*. Thus, $\delta\phi$ represents an admissible variation of function $\phi(x, y)$. In general, however, the symbol $\delta()$ stands for *change* or *variation* in the quantity within the parentheses ().

Since J is a function of $\phi(x, y)$ any change in $\phi(x, y)$ will also result into a corresponding change in J. Let δJ be the corresponding change in J due to an *admissible variation*, $\delta\phi$, in the function $\phi(x, y)$ so that

$$\delta J = ½K\int_{\mathcal{R}}\left[\delta\left(\frac{\partial\phi}{\partial x}\right)^2 + \delta\left(\frac{\partial\phi}{\partial y}\right)^2\right]d\mathcal{R} + \int_{S_q}\overline{q_n}\delta\phi\, dS \qquad (9.3)$$

If $J = J(\phi)$ attains a stationary value at the argument, $\phi = \phi(x, y)$, the change in the functional, δJ, must be equal to zero for arbitrarily small admissible variations in $\phi = \phi(x, y)$. Thus, setting $\delta J = 0$ yields the following:

$$½K\int_{\mathcal{R}}\left[\delta\left(\frac{\partial\phi}{\partial x}\right)^2 + \delta\left(\frac{\partial\phi}{\partial y}\right)^2\right]d\mathcal{R} + \int_{S_q}\overline{q_n}\,\delta\phi\, dS = 0 \qquad (9.4a)$$

$$K\int_{\mathcal{R}}\left[\frac{\partial\phi}{\partial x}\delta\left(\frac{\partial\phi}{\partial x}\right) + \left(\frac{\partial\phi}{\partial y}\right)\delta\left(\frac{\partial\phi}{\partial y}\right)\right]d\mathcal{R} + \int_{S_q}\overline{q_n}\,\delta\phi\, dS = 0 \qquad (9.4b)$$

$$K\int_{\mathcal{R}}\left[\frac{\partial\phi}{\partial x}\frac{\partial(\delta\phi)}{\partial x} + \frac{\partial\phi}{\partial y}\frac{\partial(\delta\phi)}{\partial y}\right]d\mathcal{R} + \int_{S_q}\overline{q_n}\delta\phi\, dS = 0 \qquad (9.4c)$$

Equation 9.4b follows from the ordinary differential calculus of infinitesimals and Equation 9.4c from the following identity (see Appendix D for proof):

$$\delta\left(\frac{dy}{dx}\right) \equiv \frac{d}{dx}(\delta y) \qquad (9.5)$$

From differential calculus of infinitesimals, we can easily verify the following equalities (using the rule for derivative of product of two functions):

$$\frac{\partial\phi}{\partial x}\frac{\partial(\delta\phi)}{\partial x}=\frac{\partial}{\partial x}\left[\frac{\partial\phi}{\partial x}\delta\phi\right]-\frac{\partial^2\phi}{\partial x^2}\delta\phi \tag{9.6a}$$

$$\frac{\partial\phi}{\partial y}\frac{\partial(\delta\phi)}{\partial y}=\frac{\partial}{\partial y}\left[\frac{\partial\phi}{\partial y}\delta\phi\right]-\frac{\partial^2\phi}{\partial y^2}\delta\phi \tag{9.6b}$$

Substituting the aforementioned values in Equation 9.4c yields the following:

$$-K\int_{\mathcal{R}}\left[\left(\frac{\partial^2\phi}{\partial x^2}+\frac{\partial^2\phi}{\partial y^2}\right)\delta\phi\right]d\mathcal{R}+K\int_{\mathcal{R}}\left[\frac{\partial}{\partial x}\left[\frac{\partial\phi}{\partial x}\delta\phi\right]\right.$$
$$\left.+\frac{\partial}{\partial y}\left[\frac{\partial\phi}{\partial y}\delta\phi\right]\right]d\mathcal{R}+\int_{S_q}\overline{q_n}\delta\phi\, dS=0 \tag{9.7a}$$

Using Gauss's theorem, the second area integral can be converted into boundary integral as shown in the following:

$$-K\int_{\mathcal{R}}\left[\left(\frac{\partial^2\phi}{\partial x^2}+\frac{\partial^2\phi}{\partial y^2}\right)\delta\phi\right]d\mathcal{R}+K\int_{S}\left[\left[\frac{\partial\phi}{\partial x}\delta\phi\right]n_x\right.$$
$$\left.+\left[\frac{\partial\phi}{\partial y}\delta\phi\right]n_y\right]dS+\int_{S_q}\overline{q_n}\delta\phi\, dS=0 \tag{9.7b}$$

$$-K\int_{\mathcal{R}}\left[\left(\frac{\partial^2\phi}{\partial x^2}+\frac{\partial^2\phi}{\partial y^2}\right)\delta\phi\right]d\mathcal{R}+K\int_{S}\left[\left[\frac{\partial\phi}{\partial x}\right]n_x\right.$$
$$\left.+\left[\frac{\partial\phi}{\partial y}\right]n_y\right]\delta\phi\, dS+\int_{S_q}\overline{q_n}\delta\phi\, dS=0 \tag{9.7c}$$

where n_x, n_y are the components of unit outward normal to S. Since ϕ is the specified boundary condition on S_ϕ, we require that the *admissible variation* should be zero on this part of the boundary—that is, $\delta\phi=0$ on S_ϕ. Thus, the last two boundary integrals in Equation 9.7c can be combined to obtain the following result:

$$-K\int_{\mathcal{R}}\left[\left(\frac{\partial^2\phi}{\partial x^2}+\frac{\partial^2\phi}{\partial y^2}\right)\delta\phi\right]d\mathcal{R}+\int_{S_q}\left[K\left(\left[\frac{\partial\phi}{\partial x}\right]n_x+\left[\frac{\partial\phi}{\partial y}\right]n_y\right)+\overline{q_n}\right]$$
$$\times\delta\phi\, dS=0 \tag{9.7d}$$

The aforementioned equation remains true for all arbitrarily small *admissible variations* in $\phi=\phi(x, y)$. Since $\delta\phi$ is arbitrary in $\mathcal{R}$ and also on S_q, it follows from the aforementioned equation that

$$\frac{\partial^2\phi}{\partial x^2}+\frac{\partial^2\phi}{\partial y^2}=0 \quad \text{on } \mathcal{R} \tag{9.8a}$$

$$-K\left(\left[\frac{\partial\phi}{\partial x}\right]n_x+\left[\frac{\partial\phi}{\partial y}\right]n_y\right)=\overline{q_n} \quad \text{on } S_q \tag{9.8b}$$

Thus, we come to the conclusion that when J attains a stationary value—that is, $\delta J=0$, the argument function $\phi(x, y)$, for which J attains a stationary value, is also the solution to the boundary-value problem, because, according to Equation 9.8a, $\phi(x, y)$ satisfies the Laplace equation at every point of $\mathcal{R}$ and also meets the boundary condition on S_q, according to Equation 9.8b. What about the boundary condition on S_ϕ? We have satisfied that boundary condition *by requiring that the admissible variations should be zero on* S_ϕ. We have used this fact in arriving at Equation 9.7d from 9.7c. Thus, the function $\phi=\phi(x, y)$, which gives the stationary value to the functional J, is also the solution to the boundary-value problem.

A few words about the boundary conditions are in order. First, we see, from Equation 9.8b, that the specified function $\overline{q_n}$ on S_q boundary indeed represents the outward normal flux; second, we force compliance with boundary condition on S_ϕ by requiring that the admissible variation should be zero on this boundary. In the parlance of calculus of variations, we call the boundary condition on S_ϕ as the *forced boundary condition* and the boundary condition on S_q as *the natural boundary condition.*

9.2 FINITE ELEMENT FORMULATION FOR THE LAPLACE EQUATION

In finite element method, we directly work with the variational formulation of the boundary-value problem. For this purpose, Equation 9.4b can best be written in matrix form as shown in the following:

$$\delta J=\int_{\mathcal{R}}\begin{Bmatrix}\delta\left(\dfrac{\partial\phi}{\partial x}\right)\\ \delta\left(\dfrac{\partial\phi}{\partial y}\right)\end{Bmatrix}^T\begin{bmatrix}K & 0\\ 0 & K\end{bmatrix}\begin{Bmatrix}\left(\dfrac{\partial\phi}{\partial x}\right)\\ \left(\dfrac{\partial\phi}{\partial y}\right)\end{Bmatrix}d\mathcal{R}+\int_{S_q}\{\delta\phi\}^T\{\overline{q_n}\}\,dS=0 \tag{9.9}$$

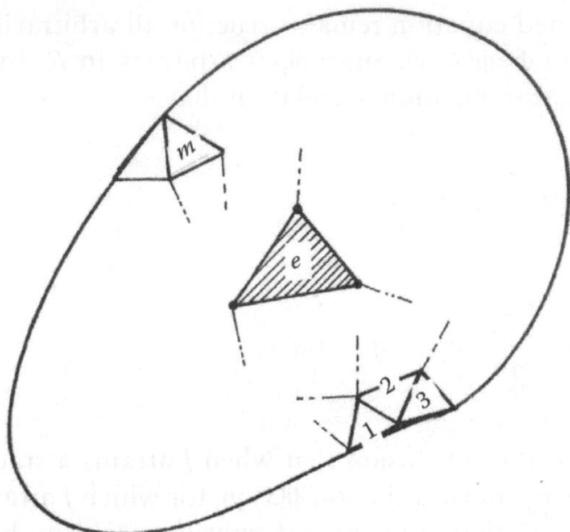

Figure 9.2 Discretization of flow region $\mathcal{R}$.

where $\{\}^T$ denotes the transpose of the column matrix $\{\}$. We divide the flow region $\mathcal{R}$ into a number (m) of subregions, as shown in Figure 9.2. For simplicity, we assume each subregion to be of triangular shape. Let δJ_e denote the variation in J due to an admissible variation of ϕ in a typical element e. Then, δJ_e can be found by evaluating the two integrals given in Equation 9.9 over a typical element e. Let us break this variation in J into two parts:

$$\delta J_e = \delta J_e^{int} + \delta J_e^{ext} \tag{9.10a}$$

The first part is given by

$$\delta J_e^{int} = \int_{\mathcal{R}_e} \begin{Bmatrix} \delta\left(\dfrac{\partial \phi}{\partial x}\right) \\ \delta\left(\dfrac{\partial \phi}{\partial y}\right) \end{Bmatrix}^T \begin{bmatrix} K & 0 \\ 0 & K \end{bmatrix} \begin{Bmatrix} \left(\dfrac{\partial \phi}{\partial x}\right) \\ \left(\dfrac{\partial \phi}{\partial y}\right) \end{Bmatrix} d\mathcal{R} \tag{9.10b}$$

where $\mathcal{R}_e$ denotes the domain of integration for a typical element, e, and the second part by

$$\delta J_e^{ext} = \int_{S_{qe}} \{\delta\phi\}^T \{\overline{q_n}\} dS \tag{9.10c}$$

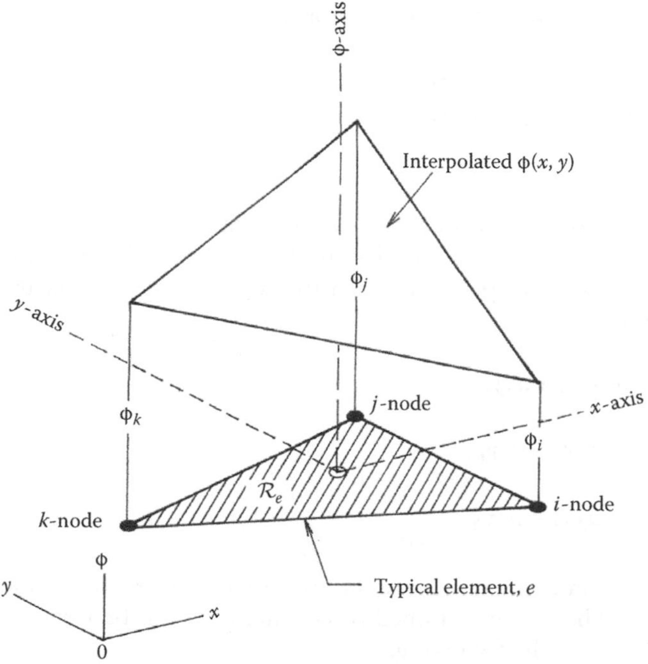

Figure 9.3 A general linear behavior of $\phi(x, y)$ in subregion, $\mathcal{R}_e$.

In the aforementioned equation, S_{q_e} denotes the boundary of element e where $\overline{q_n}$ has been specified. The boundary integral over S_{q_e} in Equation 9.10c for a typical element located in the interior of $\mathcal{R}$ is zero unless a part of the element boundary coincides with a part of S_q of region $\mathcal{R}$. Presently, we shall postpone a discussion on surface integral and, instead, focus our attention on the details of contribution to δJ_e by the area integral, given in Equation 9.10b. For this purpose, a reference to Figure 9.3 will be handy for further discussion.

We assume a priori a general linear behavior of the unknown function $\phi = \phi(x, y)$ within the element e. To describe this linear behavior of ϕ mathematically, we embed a local x, y, ϕ coordinate system at the centroid of the typical element e, as shown in Figure 9.3. The choice of a local coordinate system is arbitrary; however, we choose the centroid for the origin and keep x- and y-axis parallel to the so-called global coordinate system (x-0-y in the figure). Any linear behavior of $\phi = \phi(x, y)$ in terms of x and y can be written as

$$\phi(x,y) = a_1 + a_2 x + a_3 y \tag{9.11a}$$

which can also be written in matrix form as

$$\phi(x,y) = \lfloor 1 \quad x \quad y \rfloor \begin{Bmatrix} a_1 \\ a_2 \\ a_3 \end{Bmatrix} \tag{9.11b}$$

where a_1, a_2, a_3 are the unknown coefficients and $\lfloor\ \rfloor$ represents a row matrix (or vector). If ϕ_i, ϕ_j, ϕ_k denote the values of the function ϕ at the three nodes i, j, k, respectively, then the following three linear equations must be true:

$$\phi_i = a_1 + a_2 x_i + a_3 y_i \tag{9.12a}$$

$$\phi_j = a_1 + a_2 x_j + a_3 y_j \tag{9.12b}$$

$$\phi_k = a_1 + a_2 x_k + a_3 y_k \tag{9.12c}$$

where (x_i, y_i), (x_j, y_j), (x_k, y_k) are the coordinates of the three nodes, i, j, k, respectively. The aforementioned three equations can be written in matrix form as shown in the following:

$$\begin{Bmatrix} \phi_i \\ \phi_j \\ \phi_k \end{Bmatrix} = \begin{bmatrix} 1 & x_i & y_i \\ 1 & x_j & y_j \\ 1 & x_k & y_k \end{bmatrix} \begin{Bmatrix} a_1 \\ a_2 \\ a_3 \end{Bmatrix} \tag{9.12d}$$

which in condensed symbolic notation may be written as

$$\{\phi\} = [A]\{a\} \tag{9.13}$$

where

$$\{\phi\} \equiv \begin{Bmatrix} \phi_i \\ \phi_j \\ \phi_k \end{Bmatrix}, \quad [A] \equiv \begin{bmatrix} 1 & x_i & y_i \\ 1 & x_j & y_j \\ 1 & x_k & y_k \end{bmatrix}, \quad \text{and} \quad \{a\} \equiv \begin{Bmatrix} a_1 \\ a_2 \\ a_3 \end{Bmatrix}$$

Equation 9.11b can be written in condensed symbolic notation as (Figure 9.3)

$$\phi(x,y)\big|_{\text{over a typical element } e} = \lfloor 1 \quad x \quad y \rfloor [A]^{-1}\{\phi\} \tag{9.14}$$

It may be emphasized that on the right-hand side in the aforementioned equation, only the row vector $\lfloor 1 \quad x \quad y \rfloor$ depends on the spatial

Table 9.1 Matrix of nodal coordinates [A] and its inverse $[A]^{-1}$

$$[A]=\begin{bmatrix}1 & x_i & y_i\\ 1 & x_j & y_j\\ 1 & x_k & y_k\end{bmatrix}$$

The inverse matrix is given as follows:

$$[A]^{-1}=\frac{1}{2\Delta_{ijk}}\begin{bmatrix}x_jy_k-x_ky_j & x_ky_i-x_iy_k & x_iy_j-x_jy_i\\ y_j-y_k & y_k-y_i & y_i-y_j\\ x_k-x_j & x_i-x_k & x_j-x_i\end{bmatrix}$$

where $2\Delta_{ijk}$ denotes the determinant of [A] matrix, or twice the area of the triangular element, with nodes *i*, *j*, *k*.

coordinates *x* and *y*. The inverse matrix $[A]^{-1}$ contains only the coordinates of the three vertices, *i*, *j*, *k*, of the element *e*. The elements of this matrix are shown in Table 9.1. Thus, taking the partial derivatives of both sides of Equation 9.14 while treating the matrices $[A]^{-1}$ and $\{\phi\}$ as constants yields the following:

$$\frac{\partial\phi}{\partial x}=\lfloor 0 \quad 1 \quad 0\rfloor[A]^{-1}\{\phi\} \tag{9.15a}$$

$$\frac{\partial\phi}{\partial y}=\lfloor 0 \quad 0 \quad 1\rfloor[A]^{-1}\{\phi\} \tag{9.15b}$$

or in a combined matrix form

$$\begin{Bmatrix}\left(\dfrac{\partial\phi}{\partial x}\right)\\ \left(\dfrac{\partial\phi}{\partial y}\right)\end{Bmatrix}=\begin{bmatrix}0 & 1 & 0\\ 0 & 0 & 1\end{bmatrix}[A]^{-1}\{\phi\} \tag{9.15c}$$

From the aforementioned equation, it follows

$$\begin{Bmatrix}\delta\left(\dfrac{\partial\phi}{\partial x}\right)\\ \delta\left(\dfrac{\partial\phi}{\partial y}\right)\end{Bmatrix}=\begin{bmatrix}0 & 1 & 0\\ 0 & 0 & 1\end{bmatrix}[A]^{-1}\{\delta\phi\} \tag{9.16}$$

where

$$\{\delta\phi\} \equiv \begin{Bmatrix} \delta\phi_i \\ \delta\phi_j \\ \delta\phi_k \end{Bmatrix} \tag{9.17}$$

In the aforementioned identity, the elements of the vector on the right-hand side represent the variations in the nodal values of the function ϕ at i-, j-, and k-nodes, respectively, as shown in Figure 9.4.

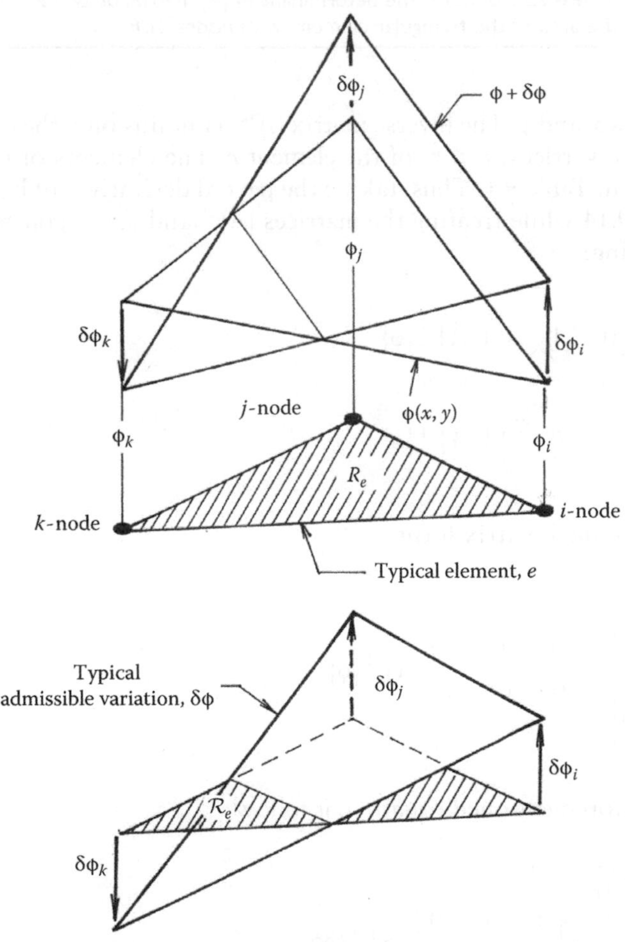

Figure 9.4 Definition sketch for the variation, $\delta\phi(x, y)$, over a typical element, e.

Now, the area integral

$$\int_{\mathcal{R}} \begin{Bmatrix} \delta\left(\dfrac{\partial\phi}{\partial x}\right) \\ \delta\left(\dfrac{\partial\phi}{\partial y}\right) \end{Bmatrix}^T \begin{bmatrix} K & 0 \\ 0 & K \end{bmatrix} \begin{Bmatrix} \left(\dfrac{\partial\phi}{\partial x}\right) \\ \left(\dfrac{\partial\phi}{\partial y}\right) \end{Bmatrix} d\mathcal{R}$$

over a typical element e can be found as shown in the following:

$$\int_{\mathcal{R}_e} \{\delta\phi\}^T \left[[A]^{-1}\right]^T \begin{bmatrix} 0 & 1 & 0 \\ 0 & 0 & 1 \end{bmatrix}^T \begin{bmatrix} K & 0 \\ 0 & K \end{bmatrix} \begin{bmatrix} 0 & 1 & 0 \\ 0 & 0 & 1 \end{bmatrix} [A]^{-1} \{\phi\} d\mathcal{R} \tag{9.18}$$

where $\mathcal{R}_e$ denotes the domain of integration for a typical element, e. The aforementioned integral can also be written as

$$\{\delta\phi\}^T \int_{\mathcal{R}_e} \left[[A]^{-1}\right]^T \begin{bmatrix} 0 & 0 & 0 \\ 0 & K & 0 \\ 0 & 0 & K \end{bmatrix} [A]^{-1} d\mathcal{R} \{\phi\} \tag{9.19}$$

because the matrices $\{\delta\phi\}^T$ and $\{\phi\}$ do not depend on the dummy variable of integration, $d\mathcal{R}$, they can be taken out of the integral sign. The integrand in the aforementioned integral represents a (3×3) matrix whose elements are constants, independent of the variable of integration. After integration over a typical element, the aforementioned expression can be represented as

$$\begin{Bmatrix} \delta\phi_i \\ \delta\phi_j \\ \delta\phi_k \end{Bmatrix}^T \begin{bmatrix} SK(1,1) & SK(1,2) & SK(1,3) \\ SK(2,1) & SK(2,2) & SK(2,3) \\ SK(3,1) & SK(3,2) & SK(3,3) \end{bmatrix} \begin{Bmatrix} \phi_i \\ \phi_j \\ \phi_k \end{Bmatrix}$$

or

$$\begin{Bmatrix} \delta\phi_i \\ \delta\phi_j \\ \delta\phi_k \end{Bmatrix}^T \begin{bmatrix} SK(i,i) & SK(i,j) & SK(i,k) \\ SK(j,i) & SK(j,j) & SK(j,k) \\ SK(k,i) & SK(k,j) & SK(k,k) \end{bmatrix} \begin{Bmatrix} \phi_i \\ \phi_j \\ \phi_k \end{Bmatrix}$$

or in condensed symbolic notation as

$$\{\delta\phi\}_e^T [SK]_e \{\phi\}_e \tag{9.20}$$

In the aforementioned expression, subscript e refers to the typical element e. The matrix $[SK]_e$ is symmetric, and in the parlance of the finite element method, this matrix is referred to as the *element stiffness matrix*. It has the dimension (3 × 3)—that is, it consists of three rows and three columns. The elements of the stiffness matrix are given in Table 9.2, in terms of the coordinates of vertices of the triangular region $\mathcal{R}_e$ and the coefficient of permeability K of the medium.

With regard to a typical element e, we are at a point where we can state that a part of the contribution, J_e^{int}, to δJ_e is given by the following equation:

$$\delta J_e^{int} = \{\delta\phi\}_e^T [SK]_e \{\phi\}_e \tag{9.21}$$

Table 9.2 Details of element stiffness matrix [*SK*]

$$SK(1,1) = SK(i,i) = \frac{K}{4\Delta_{ijk}}\left[(y_j - y_k)(y_j - y_k) + (x_k - x_j)(x_k - x_j)\right]$$

$$SK(2,1) = SK(j,i) = \frac{K}{4\Delta_{ijk}}\left[(y_k - y_i)(y_j - y_k) + (x_i - x_k)(x_k - x_j)\right]$$

$$SK(3,1) = SK(k,i) = \frac{K}{4\Delta_{ijk}}\left[(y_i - y_j)(y_j - y_k) + (x_j - x_i)(x_k - x_j)\right]$$

$$SK(1,2) = SK(i,j) = \frac{K}{4\Delta_{ijk}}\left[(y_j - y_k)(y_k - y_i) + (x_k - x_j)(x_i - x_k)\right]$$

$$SK(2,2) = SK(j,j) = \frac{K}{4\Delta_{ijk}}\left[(y_k - y_i)(y_k - y_i) + (x_i - x_k)(x_i - x_k)\right]$$

$$SK(3,2) = SK(k,j) = \frac{K}{4\Delta_{ijk}}\left[(y_i - y_j)(y_k - y_i) + (x_j - x_i)(x_i - x_k)\right]$$

$$SK(1,3) = SK(i,k) = \frac{K}{4\Delta_{ijk}}\left[(y_j - y_k)(y_i - y_j) + (x_k - x_j)(x_j - x_i)\right]$$

$$SK(2,3) = SK(j,k) = \frac{K}{4\Delta_{ijk}}\left[(y_k - y_i)(y_i - y_j) + (x_i - x_k)(x_j - x_i)\right]$$

$$SK(3,3) = SK(k,k) = \frac{K}{4\Delta_{ijk}}\left[(y_i - y_j)(y_i - y_j) + (x_j - x_i)(x_j - x_i)\right]$$

Note: Δ_{ijk} denotes the area of the triangle with vertices i, j, k. Matrix [*SK*] is symmetric.

We, now, naturally divert our attention to the remaining part, J_e^{ext}, of the contribution to δJ_e. This issue is the subject matter of the proceeding paragraphs.

As mentioned briefly in the foregoing, the contribution

$$\delta J_e^{ext} = \int_{S_{qe}} \{\delta\phi\}^T \{\overline{q_n}\} dS \tag{9.10c}$$

by a typical element completely located in the interior of the flow region $\mathcal{R}$ is zero, because the element boundary S_{q_e} where the flux is specified vanishes. Only when the element happens to embrace a part of the boundary, S_q, of the main flow domain, $\mathcal{R}$, there may be a nonzero contribution to δJ_e. Such a situation is shown in Figure 9.5.

With regard to Figure 9.5, it is seen that the side $i\ j$ of the element e embraces a part of the boundary, S_q, where $\overline{q_n}$ is specified. Since the interpolated variation, $\delta\phi(x, y)$, is linear within a typical element, it is also linear on the element boundary. Thus,

$$\delta\phi(s) = \lfloor 1 \quad s \rfloor \begin{Bmatrix} a_1 \\ a_2 \end{Bmatrix} \tag{9.22}$$

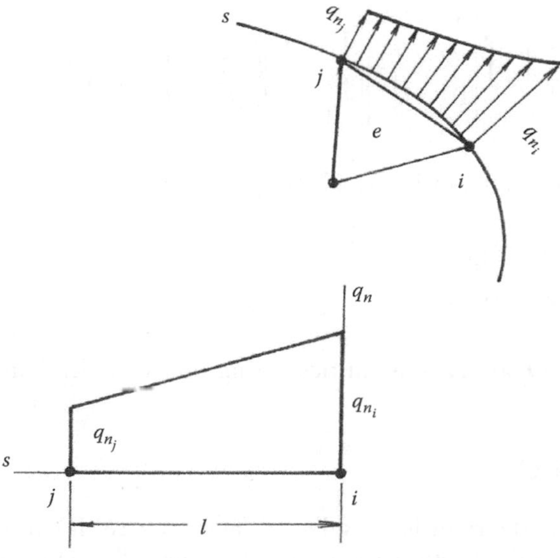

Figure 9.5 Nomenclature for equivalent groundwater flux at the nodes.

where s denotes the coordinate of a generic point along element boundary $i\,j$, as shown in the figure. If the s-coordinates of nodes i and j are $s=0$ and $s=l$, respectively, then the following equation is true

$$\begin{Bmatrix} \delta\phi_i \\ \delta\phi_j \end{Bmatrix} = \begin{bmatrix} 1 & 0 \\ 1 & l \end{bmatrix} \begin{Bmatrix} a_1 \\ a_2 \end{Bmatrix} \tag{9.23}$$

or combining the last two equation yields the following:

$$\delta\phi(s) = \lfloor 1 \quad s \rfloor \begin{bmatrix} 1 & 0 \\ \dfrac{-1}{l} & \dfrac{1}{l} \end{bmatrix} \begin{Bmatrix} \delta\phi_i \\ \delta\phi_j \end{Bmatrix} \tag{9.24}$$

If we assume a linear variation of $\overline{q_n}$ with s, we can write, similar to Equation 9.24, the following equation:

$$\overline{q_n}(s) = \lfloor 1 \quad s \rfloor \begin{bmatrix} 1 & 0 \\ \dfrac{-1}{l} & \dfrac{1}{l} \end{bmatrix} \begin{Bmatrix} q_{n_i} \\ q_{n_j} \end{Bmatrix} \tag{9.25}$$

where q_{n_i} and q_{n_j} denote, respectively, the values of normal flux at nodes i and j. Substituting Equations 9.24 and 9.25 into Equation 9.10c yields

$$\delta J_e^{ext} = \begin{Bmatrix} \delta\phi_i \\ \delta\phi_j \end{Bmatrix}^T \int_0^l \begin{bmatrix} 1 & \dfrac{-1}{l} \\ 0 & \dfrac{1}{l} \end{bmatrix} \begin{bmatrix} 1 & s \\ s & s^2 \end{bmatrix} \begin{bmatrix} 1 & 0 \\ \dfrac{-1}{l} & \dfrac{1}{l} \end{bmatrix} \begin{Bmatrix} q_{n_i} \\ q_{n_j} \end{Bmatrix} ds \tag{9.26}$$

which on definite integration yields the following:

$$\delta J_e^{ext} = \begin{Bmatrix} \delta\phi_i \\ \delta\phi_j \end{Bmatrix}^T \begin{Bmatrix} Q_i \\ Q_j \end{Bmatrix} \tag{9.27}$$

where Q_i and Q_j are two quantities acting on the nodes i and j such that the product

$$\delta\phi_i Q_i + \delta\phi_j Q_j \tag{9.28}$$

truly represents the right-hand side of Equation 9.26. It can be shown (by actual matrix multiplication and performing the definite integration shown in Equation 9.26) that

$$Q_i = \left(\frac{l}{3}\right)q_{n_i} + \left(\frac{l}{6}\right)q_{n_j} \tag{9.29a}$$

$$Q_j = \left(\frac{l}{6}\right)q_{n_i} + \left(\frac{l}{3}\right)q_{n_j} \tag{9.29b}$$

It is also clear that Q_i and Q_j represent the equivalent groundwater flows at the two nodes *i* and *j*, respectively. Furthermore, from our knowledge of statics of structures, we can recognize that Q_i and Q_j are similar to the so-called statically equivalent loads acting at the two nodes. These loads are equivalent to distributed trapezoidal loading with intensity of loading q_{n_i} and q_{n_j} at the two nodes *i* and *j*, respectively. This equivalency between the point loads and the transversely distributed loading is shown in Figure 9.6. When $q_{n_i} = q_{n_j}$, the equivalent point loads, Q_i and Q_j, reduce to the case of uniformly distributed loading. Likewise, when either q_{n_i} or q_{n_j} vanishes, the equivalent loads, Q_i and Q_j, reduce to the case of triangularly distributed loading (Figure 9.6).

It is now clear that the variation in J due to admissible variation in $\phi(x, y)$ over a typical element e (including the interior region as well as the exterior boundary) is given by the following expression:

$$\delta J_e = \delta J_e^{int} + \delta J_e^{ext} = \{\delta\phi\}_e^T [SK]_e \{\phi\}_e + \{\delta\phi\}_e^T \{Q\}_e \tag{9.30}$$

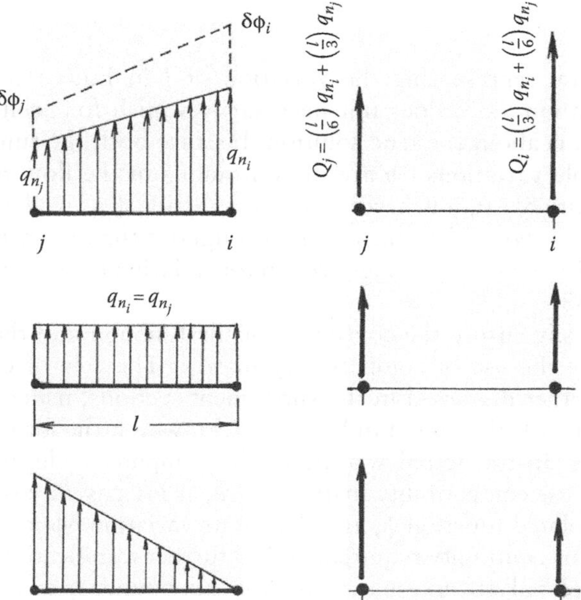

Figure 9.6 Equivalent groundwater flux at the two nodes *i* and *j*.

where

$$\{Q\}_e \equiv \begin{Bmatrix} Q_i \\ Q_j \\ Q_k \end{Bmatrix}_e \tag{9.31}$$

Thus, the total variation in J is given by the following sum:

$$\delta J = \sum_{e=1}^{e=m} \{\delta\phi\}_e^T [SK]_e \{\phi\}_e + \{\delta\phi\}_e^T \{Q\}_e \tag{9.32}$$

The aforementioned sum can also be written as

$$\delta J = \begin{Bmatrix} \{\delta\phi\}_1 \\ \{\delta\phi\}_2 \\ \vdots \\ \{\delta\phi\}_m \end{Bmatrix}^T \left(\begin{bmatrix} [SK]_1 & 0 & \cdots & 0 \\ 0 & [SK]_2 & & \\ \vdots & & \ddots & \\ 0 & & & [SK]_m \end{bmatrix} \begin{Bmatrix} \{\phi\}_1 \\ \{\phi\}_2 \\ \vdots \\ \{\phi\}_m \end{Bmatrix} + \begin{Bmatrix} \{Q\}_1 \\ \{Q\}_2 \\ \vdots \\ \{Q\}_m \end{Bmatrix} \right) \tag{9.33}$$

The total sum representing the variation of J in Equation 9.32, or in matrix Equation 9.33, does not necessarily vanish for arbitrarily small variations in ϕ about the true solution, because both the function ϕ and the admissible variations $\delta\phi$ are discontinuous in the flow region $\mathcal{R}$. In order to ensure $\delta J = 0$, it is necessary that function ϕ as well as the *admissible variation should be continuous* throughout the region $\mathcal{R}$. This continuity requirement has been ignored in formulating the sum in Equations 9.32 and 9.33.

We can easily ensure the continuity of the function and the admissible variations by the use of compatibility matrix. The role of compatibility matrix is further discussed in the subsequent sections, using the two elements, e and $e+1$, illustrated in Figure 9.7. However, the idea is more general in scope. In the actual writing of the computer code, it is not even necessary to use compatibility matrix *per se*, as long as we ensure continuity of interpolated function ϕ, and that of its variation $\delta\phi$, throughout the region $\mathcal{R}$. This continuity requirement has further ramifications during the assembly of global stiffness matrix and the flux boundary conditions, as we shall see later. These issues are discussed in some details in the following sections.

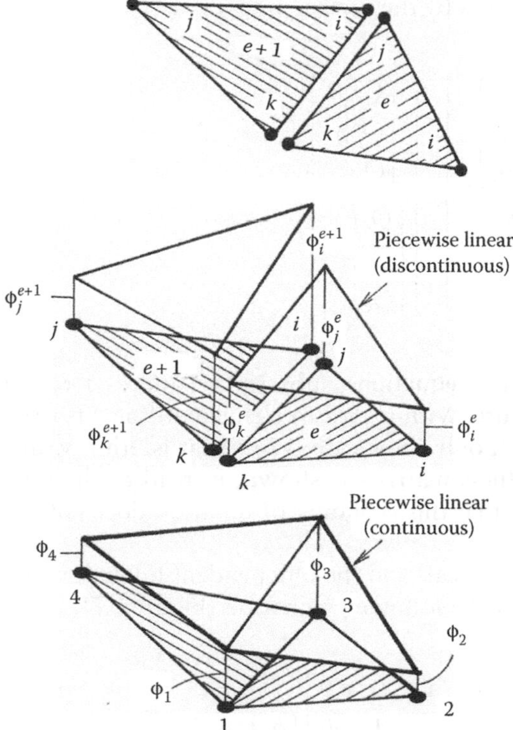

Figure 9.7 Compatibility matrix and its role.

9.2.1 Compatibility matrix, continuity of piezometric head, and related issues

For the two elements e and $e+1$ shown in Figure 9.7, we can write detailed expressions for the contributions by the area and surface integrals separately. The contribution by the area integral can be written as

$$\delta J^{int} = \begin{Bmatrix} \begin{Bmatrix} \delta\phi_i \\ \delta\phi_j \\ \delta\phi_k \end{Bmatrix}_e \\ \begin{Bmatrix} \delta\phi_i \\ \delta\phi_j \\ \delta\phi_k \end{Bmatrix}_{e+1} \end{Bmatrix}^T \begin{bmatrix} \begin{bmatrix} ii & ij & ik \\ ji & jj & jk \\ ki & kj & kk \end{bmatrix}_e & \\ & \begin{bmatrix} ii & ij & ik \\ ji & jj & jk \\ ki & kj & kk \end{bmatrix}_{e+1} \end{bmatrix} \begin{Bmatrix} \begin{Bmatrix} \phi_i \\ \phi_j \\ \phi_k \end{Bmatrix}_e \\ \begin{Bmatrix} \phi_i \\ \phi_j \\ \phi_k \end{Bmatrix}_{e+1} \end{Bmatrix} \tag{9.34}$$

and the contribution by the surface integral as

$$\delta J^{ext} = \begin{Bmatrix} \begin{Bmatrix} \delta\phi_i \\ \delta\phi_j \\ \delta\phi_k \end{Bmatrix}_e \\ \begin{Bmatrix} \delta\phi_i \\ \delta\phi_j \\ \delta\phi_k \end{Bmatrix}_{e+1} \end{Bmatrix}^T \begin{Bmatrix} \begin{Bmatrix} Q_i \\ Q_j \\ Q_k \end{Bmatrix}_e \\ \begin{Bmatrix} Q_i \\ Q_j \\ Q_k \end{Bmatrix}_{e+1} \end{Bmatrix} \tag{9.35}$$

In the previous two equations, subscripts e and $e+1$ refer to the elements shown in the figure. With regard to element stiffness matrices, $[\ ]_e$ and $[\ ]_{e+1}$ in Equation 9.34, only the location $i\ j$—that is, ith row and jth column of the element stiffness matrix—is shown, instead of the full designation $SK(i, j)$. It is hoped that this economy of notation does not cause unnecessary confusion.

The dependent (local) and the independent (global) vectors of nodal values are related by the following equation (Figure 9.7):

$$\begin{Bmatrix} \begin{Bmatrix} \phi_i \\ \phi_j \\ \phi_k \end{Bmatrix}_e \\ \begin{Bmatrix} \phi_i \\ \phi_j \\ \phi_k \end{Bmatrix}_{e+1} \end{Bmatrix} = \begin{bmatrix} 0 & 1 & 0 & 0 \\ 0 & 0 & 1 & 0 \\ 1 & 0 & 0 & 0 \\ 0 & 0 & 1 & 0 \\ 0 & 0 & 0 & 1 \\ 1 & 0 & 0 & 0 \end{bmatrix} \begin{Bmatrix} \phi_1 \\ \phi_2 \\ \phi_3 \\ \phi_4 \end{Bmatrix} \tag{9.36}$$

The aforementioned relationship is valid only if the interpolated piecewise linear ϕ is continuous across interelement boundaries. It is convenient for this discussion to write Equation 9.36 using condensed symbolic notations as

$$\begin{Bmatrix} \{\phi\}_e \\ \{\phi\}_{e+1} \end{Bmatrix} = [C]\{\phi\} \tag{9.37}$$

where the rectangular matrix $[C]$ is referred to as the *compatibility matrix* and it transforms the vector consisting of global nodal values (like $\phi_1, \phi_2, \ldots$ etc.) into a vector consisting of local nodal values (like $\phi_i^e, \phi_j^e, \ldots$ etc.). Taking the transpose of the aforementioned equation yields

$$\begin{Bmatrix} \{\phi\}_e \\ \{\phi\}_{e+1} \end{Bmatrix}^T = \{\phi\}^T [C]^T \tag{9.38a}$$

From Equation 9.38a, it follows that the variations in local and global nodal values are related by the following equation:

$$\begin{Bmatrix} \{\delta\phi\}_e \\ \{\delta\phi\}_{e+1} \end{Bmatrix}^T = \{\delta\phi\}^T [C]^T \tag{9.38b}$$

For the two elements shown in Figure 9.7, Equation 9.34 can be written, using Equations 9.37 and 9.38b, in the following form:

$$\delta J^{int} = \{\delta\phi\}^T [C]^T \begin{bmatrix} \begin{bmatrix} ii & ij & ik \\ ji & jj & jk \\ ki & kj & kk \end{bmatrix}_e & \\ & \begin{bmatrix} ii & ij & ik \\ ji & jj & jk \\ ki & kj & kk \end{bmatrix}_{e+1} \end{bmatrix} [C]\{\phi\} \tag{9.39}$$

which after (right- and left-hand) multiplication by [C] and $[C]^T$ yields the following:

$$\delta J^{int} = \{\delta\phi\}^T \begin{bmatrix} (kk)^e + (kk)^{e+1} & (ki)^e & (kj)^e + (ki)^{e+1} & (kj)^{e+1} \\ (ik)^e & (ii)^e & (ij)^e & 0 \\ (jk)^e + (ik)^{e+1} & (ji)^e & (jj)^e + (ii)^{e+1} & (ij)^{e+1} \\ (jk)^{e+1} & 0 & (ji)^{e+1} & (jj)^{e+1} \end{bmatrix} \{\phi\} \tag{9.40}$$

Using finite element vernacular, the 4×4 matrix in the aforementioned equation is called the *global stiffness matrix* or the *assembled stiffness matrix*. In the same equation, the matrices

$$\{\phi\} = \begin{Bmatrix} \phi_1 \\ \phi_2 \\ \phi_3 \\ \phi_4 \end{Bmatrix} \quad \text{and} \quad \{\delta\phi\}^T = \begin{Bmatrix} \delta\phi_1 \\ \delta\phi_2 \\ \delta\phi_3 \\ \delta\phi_4 \end{Bmatrix}^T$$

now represent the *independent vectors*, consisting of the global nodal values and the global admissible variations. It is also evident that the interpolated function ϕ and the interpolated admissible variation $\delta\phi$, using the aforementioned vectors, *are piecewise linear as well as continuous throughout the region $\mathcal{R}$.*

The role of the compatibility matrix in obtaining the global matrix from individual element stiffness matrices can be discerned by a comparative examination of two square matrices given in Equations 9.39 and 9.40. For instance, the content of the global matrix at third row and first column (highlighted in boldface) comprises the sum of the content of the element stiffness matrix at location *jk* of element *e* with the content at location *ik* of element $e+1$. Here it is seen from Figure 9.7 that location *jk* of element *e* and the location *ik* of element $e+1$ correspond with the same location 3,1 (third row and first column) of the global matrix. Thus, the multiplication of element stiffness matrices with the compatibility matrix simply establishes this correspondence and results in appropriate addition. This correspondence can be supplied by simple mapping of the local node numbering to the global node numbering, without resorting to the construction and subsequent multiplication by the compatibility matrix. This mapping is generally provided as a part of input data.

We, now, revert to the task of determining δJ^{ext}. According to Equation 9.35, this contribution for the two elements shown in Figure 9.7 is given by

$$\delta J^{ext} = \begin{Bmatrix} \begin{Bmatrix} \delta\phi_i \\ \delta\phi_j \\ \delta\phi_k \end{Bmatrix}_e \\ \begin{Bmatrix} \delta\phi_i \\ \delta\phi_j \\ \delta\phi_k \end{Bmatrix}_{e+1} \end{Bmatrix}^T \begin{Bmatrix} \begin{Bmatrix} Q_i \\ Q_j \\ Q_k \end{Bmatrix}_e \\ \begin{Bmatrix} Q_i \\ Q_j \\ Q_k \end{Bmatrix}_{e+1} \end{Bmatrix} \tag{9.35}$$

which can be written, using Equation 9.38b, as follows:

$$\delta J^{ext} = \{\delta\phi\}^T [C]^T \begin{Bmatrix} \begin{Bmatrix} Q_i \\ Q_j \\ Q_k \end{Bmatrix}_e \\ \begin{Bmatrix} Q_i \\ Q_j \\ Q_k \end{Bmatrix}_{e+1} \end{Bmatrix} \tag{9.41a}$$

or

$$\delta J^{ext} = \{\delta\phi\}^T \begin{bmatrix} 0 & 0 & 1 & 0 & 0 & 1 \\ 1 & 0 & 0 & 0 & 0 & 0 \\ 0 & 1 & 0 & 1 & 0 & 0 \\ 0 & 0 & 0 & 0 & 1 & 0 \end{bmatrix} \begin{Bmatrix} \begin{Bmatrix} Q_i \\ Q_j \\ Q_k \end{Bmatrix}_e \\ \begin{Bmatrix} Q_i \\ Q_j \\ Q_k \end{Bmatrix}_{e+1} \end{Bmatrix} \tag{9.41b}$$

or

$$\delta J^{ext} = \{\delta\phi\}^T \begin{Bmatrix} Q_k^e + Q_k^{e+1} \\ Q_i^e \\ Q_j^e + Q_i^{e+1} \\ Q_j^{e+1} \end{Bmatrix} = \{\delta\phi\}^T \begin{Bmatrix} Q_1 \\ Q_2 \\ Q_3 \\ Q_4 \end{Bmatrix} \tag{9.41c}$$

Now, we can combine the two contributions to obtain

$$\delta J = \begin{Bmatrix} \delta\phi_1 \\ \delta\phi_2 \\ \delta\phi_3 \\ \delta\phi_4 \end{Bmatrix}^T \left(\begin{bmatrix} (kk)^e + (kk)^{e+1} & (ki)^e & (kj)^e + (ki)^{e+1} & (kj)^{e+1} \\ (ik)^e & (ii)^e & (ij)^e & 0 \\ (jk)^e + (ik)^{e+1} & (ji)^e & (jj)^e + (ii)^{e+1} & (ij)^{e+1} \\ (jk)^{e+1} & 0 & (ji)^{e+1} & (jj)^{e+1} \end{bmatrix} \times \begin{Bmatrix} \phi_1 \\ \phi_2 \\ \phi_3 \\ \phi_4 \end{Bmatrix} + \begin{Bmatrix} Q_1 \\ Q_2 \\ Q_3 \\ Q_4 \end{Bmatrix} \right) = 0 \tag{9.42}$$

Since the interpolated function ϕ and admissible variations are continuous, δJ must equal zero for arbitrarily small variations around the true solution. Furthermore, since the following vector

$$\begin{Bmatrix} \delta\phi_1 \\ \delta\phi_2 \\ \delta\phi_3 \\ \delta\phi_4 \end{Bmatrix}^T$$

is completely *arbitrary*, the terms inside the braces () must be equal to zero. In other words,

$$\begin{bmatrix} (kk)^e+(kk)^{e+1} & (ki)^e & (kj)^e+(ki)^{e+1} & (kj)^{e+1} \\ (ik)^e & (ii)^e & (ij)^e & 0 \\ (jk)^e+(ik)^{e+1} & (ji)^e & (jj)^e+(ii)^{e+1} & (ij)^{e+1} \\ (jk)^{e+1} & 0 & (ji)^{e+1} & (jj)^{e+1} \end{bmatrix} \begin{Bmatrix} \phi_1 \\ \phi_2 \\ \phi_3 \\ \phi_4 \end{Bmatrix} + \begin{Bmatrix} Q_1 \\ Q_2 \\ Q_3 \\ Q_4 \end{Bmatrix} = \begin{Bmatrix} 0 \\ 0 \\ 0 \\ 0 \end{Bmatrix} \tag{9.43a}$$

or

$$\begin{bmatrix} (kk)^e+(kk)^{e+1} & (ki)^e & (kj)^e+(ki)^{e+1} & (kj)^{e+1} \\ (ik)^e & (ii)^e & (ij)^e & 0 \\ (jk)^e+(ik)^{e+1} & (ji)^e & (jj)^e+(ii)^{e+1} & (ij)^{e+1} \\ (jk)^{e+1} & 0 & (ji)^{e+1} & (jj)^{e+1} \end{bmatrix} \begin{Bmatrix} \phi_1 \\ \phi_2 \\ \phi_3 \\ \phi_4 \end{Bmatrix} = - \begin{Bmatrix} Q_1 \\ Q_2 \\ Q_3 \\ Q_4 \end{Bmatrix} \tag{9.43b}$$

which represents a system of four simultaneous equations in four unknowns,

$$\begin{Bmatrix} \phi_1 \\ \phi_2 \\ \phi_3 \\ \phi_4 \end{Bmatrix}.$$

In Equation 9.43, the 4 × 4 matrix is the same global stiffness matrix given in Equation 9.40, and the right-hand vector represents the known fluxes given in Equation 9.41c. If the global stiffness matrix is invertible (which in the case of Laplace equation is), the solution can be found.

What we have discussed is essentially the finite element method in its most basic form. However, there is one more aspect of the method that needs our attention before we can actually implement the method. This aspect constitutes the subject matter of the following section.

9.2.2 Treatment of known values of ϕ at the nodes

In the foregoing, we had assumed that all nodal values in the vector

$$\begin{Bmatrix} \phi_1 \\ \phi_2 \\ \phi_3 \\ \phi_4 \end{Bmatrix}$$

were unknown. This might not be the case. There are more than one ways to handle this situation. One of the ways to handle this situation is described here with an example. Let $\phi_2 = \overline{\phi}_2$ be a known value in Equation 9.43. We adjust the finite element Equations 9.43 in the following manner:

$$\begin{bmatrix} (kk)^e + (kk)^{e+1} & 0 & (kj)^e + (ki)^{e+1} & (kj)^{e+1} \\ 0 & 1 & 0 & 0 \\ (jk)^e + (ik)^{e+1} & 0 & (jj)^e + (ii)^{e+1} & (ij)^{e+1} \\ (jk)^{e+1} & 0 & (ji)^{e+1} & (jj)^{e+1} \end{bmatrix} \begin{Bmatrix} \phi_1 \\ \phi_2 \\ \phi_3 \\ \phi_4 \end{Bmatrix}$$

$$= -\begin{Bmatrix} Q_1 \\ Q_2 \\ Q_3 \\ Q_4 \end{Bmatrix} - \overline{\phi}_2 \begin{Bmatrix} (ki)^e \\ 0 \\ (ji)^e \\ 0 \end{Bmatrix} \tag{9.44}$$

After the aforementioned adjustment to the finite element equations, we treat as if all ϕs are unknown. The adjustment can be performed in two steps: first, multiply column 2 of the global stiffness matrix by the known value $\overline{\phi}_2$ and bring the product to the right-hand side; second, replace the second equation in the set by $\phi_2 = \overline{\phi}_2$. From the perspective of computer programming, this adjustment might be the simplest to implement. Once the adjustments are made, any computer routine can be used to invert the global stiffness matrix and thus obtain the answer for the unknown nodal vector,

$$\begin{Bmatrix} \phi_1 \\ \phi_2 \\ \phi_3 \\ \phi_4 \end{Bmatrix}.$$

SUGGESTED READINGS

Akin, J. E. 1982. *Application and Implementation of Finite Element Methods*, Academic Press, Inc. Ltd., London, U.K. First paperback edition printed in 1984.

Huebner, K. H., Dewhirst, D. L., Smith, D. E., and T. G. Byron. 2001. *The Finite Element Method for Engineers*, 4th edn., John Wiley & Sons, Inc., New York.

Reddy, J. N. 2006. *An Introduction to the Finite Element Method*, 3rd edn., McGraw-Hill Co., Inc., New York.

Strack, O. D. L. 1989. *Groundwater Mechanics*, Prentice Hall, Inc., Englewood Cliffs, NJ.

Verruijt, A. 1970. *Theory of Groundwater Flow*, Macmillan and Co. Ltd, London, U.K.

Wang, H. F. and M. P. Anderson. 1982. *Introduction to Groundwater Modeling Finite Difference and Finite Element Methods*, W. H. Freeman and Company, San Francisco, CA.

Zienkiewicz, O. C. 1982. *The Finite Element Method*, 3rd edn., McGraw-Hill Co. Ltd., London, U.K.

Appendix A: Identical similarity between Figures 1.5 and 1.6

To begin with, we assume, for simplicity, that the pore space of the aquifer can be represented by the inner space of an ensemble of capillary tubes whose inner radii range continuously from zero to a maximum radius r_0, as shown in the accompanying sketch. It is further assumed that each tube is oriented vertically and has a uniform radius. It is, however, assumed that the acceleration due to gravity acts vertically downward.

Our basic hypothesis is

$$N_r \propto \frac{1}{r^2} \tag{A.1}$$

$$N_r - \frac{C}{r^2} \tag{A.2}$$

where

N_r denotes the number of capillary tubes with (inner) radius r

C is a constant of proportionality

Let us take a cross section of the ensemble of capillary tubes at a height h, which represents the rise of water in a capillary tube of radius r (Figure A.1). It is evident that at this cross section, all capillary tubes with radii less than or equal to r are filled with water. If A_w represents the total area of the cross section occupied by water, it can be obtained by the following definite integral:

$$A_w = \int_0^r N_r \pi r^2 dr = \int_0^r \frac{C}{r^2} \pi r^2 dr = C\pi \int_0^r dr = C\pi r \tag{A.3}$$

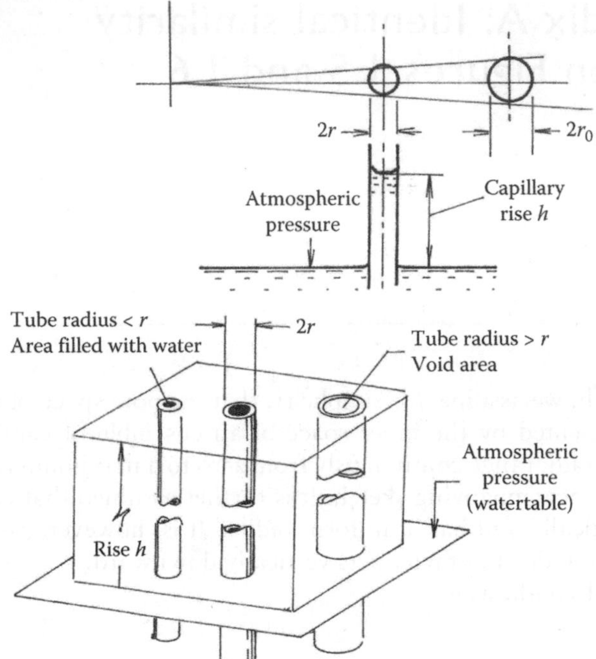

Figure A.1 Definition sketch.

Likewise, if A represents the total pore area at the cross section, it is represented by the following definite integral:

$$A = \int_0^{r_0} N_r \pi r^2 dr = C\pi r_0 \tag{A.4}$$

From the last two equations, we obtain the following:

$$\frac{A_w}{A} = \frac{r}{r_0} \tag{A.5}$$

The ratio A_w/A on the left-hand side of Equation A.5 represents the degree of saturation in fraction (as opposed to percentile ratio). It is now apparent from Equation A.5 that the normalized radius r/r_0 and the degree of saturation expressed in fraction are identical.

The normalized capillary rise h_c/h_{c0} in Figure 1.5 also represents the normalized rise of water above watertable, z/z_f. From the viewpoint of physics, both refer to the same normalized distance, only the notation has been changed. Since the abscissae and the ordinates in Figures 1.5 and 1.6 are, respectively, identical, it is evident that the two graphs in these figures are also identical.

Appendix B: Transformation of components of a vector under rotation of reference frame

In two-dimensional space, let the directed line segment OP represent a vector **x**, whose components with respect to x, y-reference frame be denoted by the ordered pair (x, y). Let another x', y'-reference frame be imbedded in this space and the components of **x** be denoted by (x', y'). For the sake of discussion, we shall speak of x', y'-reference frame as being obtained from the rotation of the x, y-reference frame through an angle θ in a counterclockwise direction (Figure B.1). With respect to the geometric construction shown in Figure B.2, we make the following observations: (1) Points x, x' represent the orthogonal projections of point P on the x- and x'-axis, respectively; likewise, points y, y' represent the orthogonal projections of point P on the y- and y'-axis, respectively; and (2) the triangles Pxx''', Pyy''', Oxx'', and Oyy'' are all right-angle triangles. From the geometric construction shown in Figure B.2, we conclude the following relationships between distances:

$$\overline{Ox'} = \overline{Ox''} + \overline{x''x'} = \overline{Ox''} + \overline{xx'''} \tag{B.1a}$$

$$x' = x\cos\theta + y\sin\theta \tag{B.1b}$$

and

$$\overline{Oy'} = \overline{Oy''} - \overline{y''y'} = \overline{Oy''} - \overline{yy'''} \tag{B.2a}$$

$$y' = y\cos\theta - x\sin\theta \tag{B.2b}$$

The last two equations can be combined into a single matrix equation, as follows:

$$\begin{Bmatrix} x' \\ y' \end{Bmatrix} = \begin{bmatrix} \cos\theta & \sin\theta \\ -\sin\theta & \cos\theta \end{bmatrix} \begin{Bmatrix} x \\ y \end{Bmatrix} \tag{B.3}$$

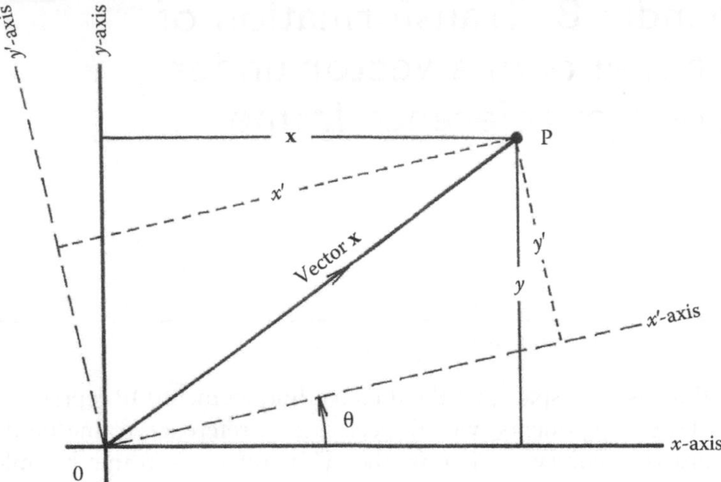

Figure B.1 Definition sketch for vector **x** and its components (x, y) and (x', y') in the two reference frames.

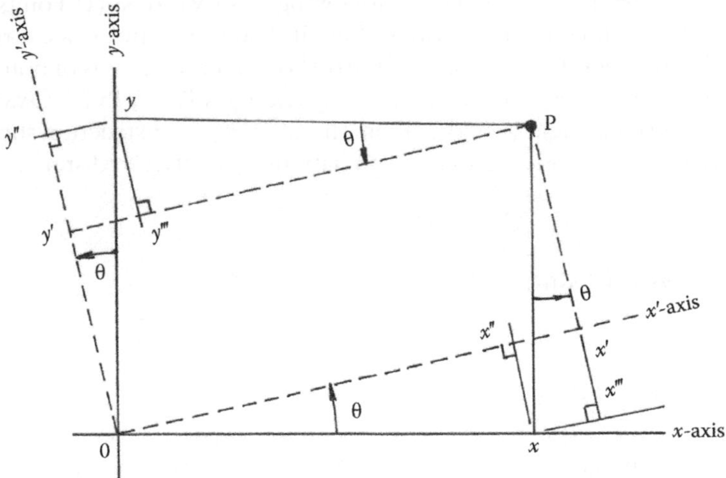

Figure B.2 Definition sketch for geometric construction.

The Equation B.3 shows how the components of a vector **x** transform when the reference frame is rotated through an angle θ in a counterclockwise direction. We emphasize that in the previous equation, *a positive value of* θ *implies a counterclockwise rotation of x, y*-reference frame *to x′, y′*-reference frame.

Appendix C: Table of well function $W(u)$

u	1.0	2.0	3.0	4.0	5.0	6.0	7.0	8.0	9.0
$\times 10^{0}$	0.219	0.049	0.013	0.0038	0.0011	0.00036	0.00012	0.000038	0.000012
$\times 10^{-1}$	1.82	1.22	0.91	0.70	0.56	0.45	0.37	0.31	0.26
$\times 10^{-2}$	4.04	3.35	2.96	2.68	2.47	2.30	2.15	2.03	1.92
$\times 10^{-3}$	6.33	5.64	5.23	4.95	4.73	4.54	4.39	4.26	4.14
$\times 10^{-4}$	8.63	7.94	7.53	7.25	7.02	6.84	6.69	6.55	6.44
$\times 10^{-5}$	10.94	10.24	9.84	9.55	9.33	9.14	8.99	8.86	8.74
$\times 10^{-6}$	13.24	12.55	12.14	11.85	11.63	11.45	11.29	11.16	11.04
$\times 10^{-7}$	15.54	14.85	14.44	14.15	13.93	13.75	13.60	13.46	13.34
$\times 10^{-8}$	17.84	17.15	16.74	16.46	16.23	16.05	15.90	15.76	15.65
$\times 10^{-9}$	20.15	19.45	19.05	18.76	18.54	18.35	18.20	18.07	17.95
$\times 10^{-10}$	22.45	21.76	21.35	21.06	20.84	20.66	20.50	20.37	20.25
$\times 10^{-11}$	24.75	24.06	23.65	23.36	23.14	22.96	22.81	22.67	22.55
$\times 10^{-12}$	27.05	26.36	25.96	25.67	25.44	25.26	25.11	24.97	24.86
$\times 10^{-13}$	29.36	28.66	28.26	27.97	27.75	27.56	27.41	27.28	27.16
$\times 10^{-14}$	31.66	30.97	30.56	30.27	30.05	29.87	29.71	29.58	29.46
$\times 10^{-15}$	33.96	33.27	32.86	32.58	32.35	32.17	32.02	31.88	31.76

Source: After Wenzel, L.K., *Method of Determining Permeability of Water-Bearing Materials with Special Reference to Discharging Well Methods*, U.S. Geological Survey Water Supply Paper 887, 1942.

SUGGESTED READING

Wenzel, L. K. 1942. *Method of Determining Permeability of Water-Bearing Materials with Special Reference to Discharging Well Methods*, U.S. Geological Survey Water Supply Paper 887.

Appendix D: Proof of the assertion $(dy/dx) = (d/dx)(\delta y)$

Let $y = y(x)$ be any function of x and δy an admissible variation in $y(x)$, as shown in Figure D.1. We know, by definition, that $\delta(dy/dx)$ stands for the change in the derivative of the function $y = y(x)$ as the function changes from $y(x)$ to $y(x) + \delta y(x)$. This change in the derivative can be written as follows (see Figure D.1):

$$\delta\left(\frac{dy}{dx}\right) = \frac{d}{dx}(y + \delta y) - \frac{d}{dx}(y)$$

$$= \frac{d}{dx}(y) + \frac{d}{dx}(\delta y) - \frac{d}{dx}(y) = \frac{d}{dx}(\delta y) \quad \text{(D.1)}$$

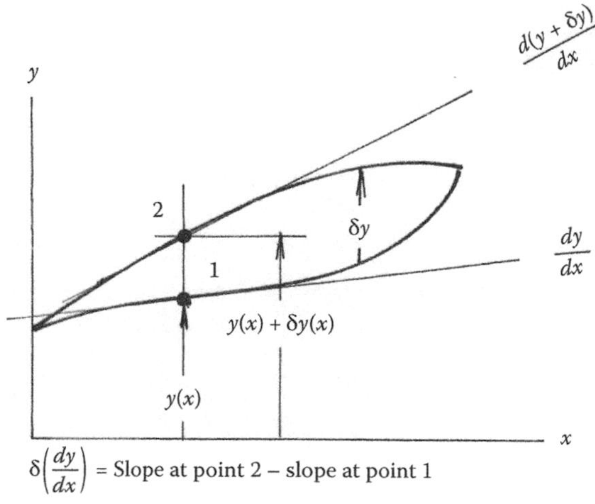

Figure D.1 Definition sketch.

In Equation D.1, the second equality follows from the differential calculus of infinitesimals. Thus,

$$\delta\left(\frac{dy}{dx}\right)=\frac{d}{dx}(\delta y) \tag{D.2}$$

Index

J

K

L

M

N

O

P

R

Printed and bound by CPI Group (UK) Ltd, Croydon, CR0 4YY
18/10/2024
01776257-0006